RADAR'S ROLE IN DEFEATING THE NAZIS

RADAR'S ROLE IN DEFEATING THE NAZIS

By Jack Nissenthall

with

BASIL SAMUELS

FRONTLINE
BOOKS

RADAR'S ROLE IN DEFEATING THE NAZIS
Foreign Recruitment in the Heer and Waffen-SS 1939-1945

First published in Great Britain in 2026
by Frontline Books
An imprint of
Pen & Sword Books Ltd
Yorkshire - Philadelphia

ISBN 9781036198244

A CIP catalogue record for this book is available from the British Library

Typeset by Lapiz Digital
Printed and bound in the UK by CPI Group (UK) Ltd,
Croydon, CR0 4YY.

Printed on paper from a sustainable source by
CPI Group (UK) Ltd, Croydon, CR0 4YY

The Publisher's authorised representative in the EU for product safety is Authorised Rep Compliance Ltd., Ground Floor, 71 Lower Baggot Street, Dublin D02 P593, Ireland.
www.arccompliance.com

For a complete list of Pen & Sword titles please contact
PEN & SWORD BOOKS LTD
47 Church Street, Barnsley, South Yorkshire, S70 2AS, England
E-mail: enquiries@pen-and-sword.co.uk
Website: www.pen-and-sword.co.uk
or
PEN & SWORD BOOKS
1950 Lawrence Rd, Havertown, PA 19083, USA
E-mail: uspen-and-sword@casematepublishers.com

CONTENTS

LIST OF ILLUSTRATIONS

FOREWORD

When my wife and I were sorting through her father's papers in 2009, after his widow, Dell, had died, we came across four ring binders containing dot matrix printed pages and placed them on a bookshelf to be looked at later.

In 2024 I eventually took the first file off the shelf and read its contents. The other three soon followed and, whilst it is a very personal view of events during the Second World War, what Jack had to tell he felt to be very important.

Jack's cousin, Hillier Wise, recalls a conversation they had about thirty-five years ago, when Jack had visited Robert Watson-Watt, who said to him, 'What we have done must be known.'

I have read many books about the course of the war and, having read my father-in-law's manuscript, realise that the contribution that Combined Operations and Radar made to the successful landings on D-Day (Operation Neptune), are barely acknowledged. This memoir attempts not to set the record straight, but to add to the narrative and fill in some of the gaps.

I have changed very little in the original manuscript, apart from correcting typos and the occasional factual error. I have also inserted several maps so that the reader can identify the places referred to in the text.

In writing this book, I felt it important that the reader should get to know Jack, and to that end the first part of this narrative is an introduction to his early life, his warmth and generous spirit and his sense of humour.

There may be certain parts of this narrative that may not be entirely or completely understood, but rather than add explanations, I decided that with access to the Internet, the reader can find further information if it is desired.

Professor R.V. Jones' foreword to Jack Nissenthall's manuscript *The Wizard War* has been taken from his biography *The Man Under the*

Radar, written by his daughter Linda Nissen Samuels, and published by Chiselbury Publishing in 2022.

I want to thank my wife, Linda, my family and friends and Pen and Sword for their support and encouragement, and of course my father-in-law, Jack, for gifting me such wonderful source material, and for being so brave and selfless.

Basil Samuels, 2026

PREAMBLE

If Jack Nissenthall had not read the book *My Life in Court* by Louis Nizer, which had been condensed in the *Reader's Digest*, most probably none of what follows would have seen the light of day. In his book, Louis Nizer had recounted some of his more unusual cases. One in particular was a libel action taken in 1955 by Quentin Reynolds, the war correspondent, journalist and author who had been accused by a rival author, Westbrook Pegler, of fabricating all his stories written in war-torn England during the Second World War. In the court case the whole outcome seemed to depend on whether Quentin Reynolds could tell the court the real name of Professor Wendell, mentioned in his book *Dress Rehearsal* (1943), the so-called scientist who had gone on the Dieppe Raid in 1942 to find out about German radar. The name of the scientist was top secret and had never been divulged; Quentin Reynolds and Louis Nizer were stumped.

It so happened that the case was reported in the *New York Times* and Jock Lawrence, a member of Mountbatten's Combined Operations staff, read about the case and decided to go to the court to see what it was all about. When Pegler's attorney repeated the question, 'Who was Professor Wendell?' Jock Lawrence decided to send a message to the judge to say that a scientist had gone on the Dieppe Raid, but that his name was secret and only known to a very few. The judge said without a name the judgement would be in favour of Pegler and that Quentin Reynolds will lose. Jock Lawrence consulted his superiors in London, who authorised him to divulge the name, but only to the judge. He wrote the name of the scientist on a piece of paper and handed it to the judge, who, having read it, screwed it up and pronounced that Quentin Reynolds had won and was awarded the largest libel award of $175,000 (about £2.1m today).

Jack, who was then living in South Africa, realised that this mad scientist they were talking about was none other than him. He wrote to Quentin Reynolds, who replied 'Where were you when I needed you?

You must write down everything you can remember about how you got into radar and what you did,' which Jack did in 1964.

The dedication at the beginning of Jack's unpublished 1964 manuscript, *The Wizard War*, reads as follows:

> This book is dedicated to '60', the radar group of the RAF without whom the Battle of Britain would have been lost. To the technical officers, RDF mechanics, Air Ministry civilians and gentlemen of universities who put aside their caps and gowns and raised the pick and shovel in those hazardous decisive days leading up to the Second World War. To this small band of men of whom many are no longer known, and whose fantastic technological achievements remain forgotten. To the young men of Britain who had the foresight to take up the defensive sciences in time of peace with an eye to the future defence of their island home, and finally to the Scot whose genius changed the course of history, Robert Watson Watt.

Jack sent his manuscript to Reginald Victor Jones, Britain's war time Chief of Air Intelligence, primarily for reassurance that what he had written was accurate. Professor Jones remembered Jack, and provided the following foreword:

> The start of radio broadcasting in the nineteen twenties brought the electronic age into the ordinary home. The principles of constructing the simple receivers of those days were within the grasp of many who had received no instruction in science; and with a little effort any man could individually listen to the world with a receiver of his own making and – if he stretched to a transmitter – speak back. He could hold personal theories about the best way to make an aerial or a loudspeaker, sometimes out of profoundly domestic items and utensils, and he could usually find, in the curious phenomena of radio propagation, facts to support his theories, however unorthodox these might be.
>
> In this atmosphere, the radio amateur came into existence and the radio shop flourished. It was fortunate for us in Britain that they did; for they provided the reserve of radio amateurs and servicemen who so effectively kept our radar and signals equipment working in the Second World War. Afterwards, our chief opponent in the radio war, General Martini (Director General of Signals of the German Air Force) told me how much he envied us this reserve. Hitler had banned amateur radio in Germany before the war because it might offer scope for anti-Nazi activities; he deprived himself of a band of men skilled in the improvisations that were necessary to keep electronic equipment in operation in the war.

Surprisingly little has been written about the men who kept our radio and radar equipment working. Although among the most skilled of jobs, it might not have appeared among the most dangerous. But it quite often was, particularly when the equipment was airborne or shipborne; and it sometimes involved the hottest of action. Harold Jordan won an immediate DSO listening in front of a German night fighter for its radio transmissions, Eric Ackerman a GM for ninety flights over the German radar defences, and William Cox an MM for his gallantry at Bruneval. Like Cox, the author of this book, Jack Nissenthall, was a Flight Sergeant Radio Mechanic in the Royal Air Force who had volunteered for an operation of unusual danger. The operation turned out to be the Dieppe Raid. One of my vivid memories is that of Flight Sergeant Nissenthall reporting his experiences to me immediately after the raid – ending with a quarter mile swim under fire to reach one of the departing ships – and his high tribute to the Canadians manning their last Bren gun on a rampart built from bodies of their dead comrades.

This book is therefore the personal story of a man who willingly went into the 'hard, savage clash' of Dieppe, spurred by patriotism and enthusiasm for electronics, and knowing that if things went wrong – which they did – he had a peculiarly slim chance of returning. That he did return was due to his great physical fitness, combined with the cheerful and resourceful courage that shows in this book. His own deeds speak for themselves; but he has generously remembered mine 'with advantages' – I only wish that I had such a tale as his to tell.

Prof. R.V. Jones
Department of Natural Philosophy
University of Aberdeen
7 April 1966

CHAPTER 1
BEGINNINGS

Jack Nissenthall was born on 19 October 1919. His father, Aaron, was born in Poland, though his mother, Annie, also from Polish Jewish stock, was born in England. Aaron was a talented master tailor and the family lived in a spacious house in Cottage Grove, Bow.

It was while at this house that young Jack devised another way of entering the building by climbing a tree and then jumping in through the window. Annie always knew he was home when she heard a loud thump.

Jack first attended Malmesbury Road School in Bow. He obviously liked school, because when his younger brother Mickey was 18 months old, Jack took him there, riding on the family dog. He thought it was time Mickey should start learning. He remembered the teachers saying Mickey was too young – he could not even talk yet. Unfazed, Jack told them that he could understand what Mickey was saying, and that he would interpret. They were not convinced, however, and contacted his mother, who by this time was frantic, thinking Mickey had been kidnapped.

Apparently, Jack was once also caught trying to set fire to the local park; nobody is sure why, although it was likely some sort of experiment.

Jack's father, Aaron, was a successful master tailor. Aaron's older brother Max Nissenthall had brought Aaron to England from Annopol, Poland, in 1909. Aaron trained as a tailor, specialising in ladies' clothes and eventually had his own business. On 11 September 1917 he married Annie Smietanski. They had four children, Marie (b. 1918), Jack (b. 1919), Michael (b. 1923) and, much later, Harold (b. 1939). According to the family, Aaron was quite a character. Married, with children, he decided to keep a goat in the back garden of their house in Cottage Grove. The neighbours were not impressed as the beast was always escaping and destroying their back gardens. That was the end of the goat.

Aaron was very proud of his adopted home and was ever conscious of the freedom offered by the people of Great Britain. He encouraged his children to be proud English citizens.

After a disastrous robbery of Aaron's premises in 1925, the family fell on hard times and had to move further into the East End, to Blythe Street, Bethnal Green. Jack had to change schools and was sent to Teesdale Street School. It was there he met Max Bernard and became a regular visitor to the Bernard family around the corner. Max Bernard, who was a little older than Jack, introduced him to basic electronics in a cupboard under the stairs, where the two boys set up a workshop and made small crystal radio sets to sell to the neighbours. Jack became totally fascinated with the new science of electronics.

It was also at the Bernards' house that he became friendly with Max's sister, Dally (who would later go by the name of Dell), who would first become his steady girlfriend throughout the Second World War, and later his wife.

Always very short of money due to the family's precarious finances, but with a great thirst for knowledge, in the holidays Jack would walk from the East End to the Science Museum in Kensington. For lunch he would eat fish and chips out of newspaper in the tunnel leading to the Underground, and then return for an afternoon session, before walking home again.

After he left Teesdale Street School in 1930, Jack attended Mansford Street School, which was essentially a technical college. There he was blessed to have an inspirational science teacher in A.H. Raines, and it was he who probably set him on course for his future career, starting him with a one-valve radio before moving on to constructing receivers under his guidance.

Jack was always looking for components for his 'hobby' and one day he cycled across London to the Columbia Gramophone Company shop in Petty France, where he met a man called Alan Blumlein (1903-42). An English electronics engineer, Blumlein was considered by many to be the twentieth century Faraday, with 128 patents for his inventions in telecommunications, sound recording, stereophonic sound, television and radar.

Blumlein was obviously impressed by Jack's enthusiasm for, at Jack's request and despite a heavy work schedule of his own, he agreed to give a talk to the Chiswick Radio Club, of which Jack was a member.

In 1931, the Columbia Gramophone Company was bought out and combined with other similar establishments to form EMI (Electrical and Musical Industries Ltd.). A brand-new factory with laboratories was built at Hayes in West London and Alan Blumlein went to work there.

In 1935, aged 16, Jack started working at EMI. Perhaps it was Alan Blumlein who persuaded him to apply, for Jack found himself working with him on sound reproduction, radio and early television. Jack assembled television sets and visited people's homes to install and maintain them. On 2 November 1936, BBC Television, as it was then known, began broadcasting regularly from Alexandra Palace to the London area.

In addition to his love of electronics, since he had first been able to wield a pair of scissors, Jack had become an avid model aircraft builder, and his father's workshop displayed quite an array of his constructions. These consisted in the main of biplanes and triplanes, both German and English, which had achieved fame in the First World War. Aeroplanes and flying had always thrilled him and once he had received his first bicycle, he never missed an opportunity of attending an air display, provided they were accessible by bicycle.

Each year a local air display was held at Hendon. This was a high point that Jack never missed and on Empire Days, he would cycle out, usually to the beautiful aerodrome at Biggin Hill. There were also odd, but excellent shows like the 1937 *Daily Express* air display at the then small civilian aerodrome at Gatwick, near Redhill. Lord Beaverbrook, whose son later became an ace pilot, had offered a prize for the best aerobatic exhibition. It was a considerable amount of money and attracted pilots from all over the Continent, as well as the American, Al Brown, who gave a wonderful demonstration of high power aerobatics. French and Belgian pilots also gave brilliant performances. The RAF, however, stole the show by cavorting all over the sky in an amazing series of aerobatics, seemingly with their aircraft tied together wingtip to wingtip.

Despite the international attendance there were no flyers from Germany, which rather disappointed the enthusiasts who had heard so much of German flying skill. For Jack, however, it was a great day as he managed to speak to the two great German generals, General Ernst Udet and General Erhard Milch of the Luftwaffe, who had come to observe. These charming men were insistent that every young man of sound eye and limbs should learn to fly. This seemed almost an obsession as they described the methods being used in Germany to train young pilots. Since this was some three years before the outbreak of the Second World War, it all sounded innocent enough. General Udet, Jack was sure, was an idealist and was quite certain that this unfortunate man had no idea to what use his carefully nurtured brainchild was to be put. He had no thought of the fire and terror they were to bring from the skies to the peoples of Europe. General Udet would later commit suicide almost halfway through the war, while General Milch, who in the 1920s created Lufthansa by amalgamating all the small German airlines, became the arch architect of the Luftwaffe. However, he fought a losing battle against Hitler and his foibles and was relieved of his post in June 1944.

In his memoir *The Wizard War*, Jack wrote the following:

> Back in London, the holiday was soon forgotten, and I was once again a tiny cog in the radio industry. Repairing and building radio and television sets, we strove to keep the demanding customers happy. To further my education, I attended for three evenings a week the Regent Street Polytechnic in the West End of London, that significant institution founded by Sir Quintin Hogg. My regular evening studies were at times supplemented by bonus short lecture courses by visiting scientific celebrities. It was during one of these lecture courses that the mysteries of the cathode ray tube, then a new industrial device, were so skilfully unfolded in great detail by Mr A.G. Parr. His brilliant lectures not only prepared me for my daily work in television, but what was more important, for the trying times ahead in radar.
>
> I had long since known that my life would be hopelessly bound to the new electronic science. Late night hours spent in study were always gratifying as I felt it was continually expanding my ability to do the work I most enjoyed. I was extremely happy as my work was also my hobby, which enabled me to build devices that made me the envy of my friends. I had a most magnificent record player, a large screen television set and I could listen, almost at will, to radio stations in practically any part of the world.
>
> I decided to learn as much as possible about radio and electronics and its potential applications in the field of national defence, realising that I could only be a very small cog in the defensive machine. Little did I then know that I and the comparatively small group of other young radical men were, by sheer hard work, to become of vital importance in the development of Britain's air defence system, already on the drawing board but sadly starved of personnel. Furthermore, that, though I did not know it at the time, I personally would be forced into a predicament where my capture alive by the Germans might have seriously affected our chances of surviving the war as yet many years away.

**UK radar stations where Jack was posted, 1939-43:
Bawdsey 1936-39, North Foreland 1940, Rosehearty 1940-41, Bolt Head 1941-43**

CHAPTER 2
INTRODUCTION TO RADAR

One day while at EMI, an RAF serviceman in civilian clothes came canvassing for men with knowledge of television and who were prepared to spend short periods working for the Air Ministry and be involved, on a more or less voluntary basis, in assisting with a top-secret new sort of radio that was being developed at Bawdsey, near Ipswich.

With the world so full of absorbing distractions in the way of entertainment, and most things being measured by the universal yardstick of pounds, shillings and pence, the visitor was finding it more than difficult to enrol assistants.

Some time later, Jack and the RAF serviceman met at a radio club meeting in Chiswick and, after further discussions, Jack decided to go along with him and help in any way he could. It was all very haphazard and for a month or two nothing happened. However, one Saturday morning, after a series of telephone calls, he found himself travelling northeast out of London to Felixstowe in a rather battered Fiat Cub with two other young men. The 70-mile trip seemed interminable. He had to compress himself into the rear of this tiny car, finding it more than difficult to unfold himself when they eventually stopped at a roadside pub for lunch. Jack had to admit rather shyly that he had never drunk beer before. Lunch was mainly dictated by their limited spending power and consisted of bread, cheese, beer and pickles. During the remainder of the journey to the coast, this repast played havoc with the internal atmosphere of the little car.

After skirting Ipswich, they made towards Felixstowe where they turned off and made for their destination, the location of which the driver appeared to know and about which he steadfastly refused to talk. Eventually they arrived at what looked like a towered country mansion in beautiful riverside grounds, where they were greeted by a middle-aged security guard, who made them state their business.

After a few words with the driver, who showed the security guard their letter of introduction, and with no further formalities, he directed them along the drive.

Leaving them standing at the front door, the driver disappeared inside and after what seemed to be an endless wait, he reappeared looking little the wiser for his visit. There was a further delay until another person appeared, Mr Airey, with a broken nose and a very loud voice. He lectured them at length on the need for secrecy and made them promise not to discuss any details of what they did or saw while working in the establishment. Needless to say, Jack was thrilled by the cloak and dagger atmosphere and mentally prepared himself to honour at all times, even unto death, the assurances he had given. They learned in time that beneath Mr Airey's gruff exterior and foghorn voice, there was a very wonderful helpful person.

For a short time, and under his guidance, they wandered around the area near the mansion, being heavily battered by the boisterous wind that swept in from the sea. After he had had enough of the wind he left them to their own devices.

There were number of huts in the vicinity, but very few people. There was certainly nothing that seemed to be remotely secret around them. Being more or less at a loss as to what to do, they hovered near a small group of men who were wrestling with a large pile of wooden spars which, Meccano style, they were trying to piece together. They offered their assistance and struggled with them for some two hours until they were finally rewarded with a rather wobbly structure some 15 feet high. Two older men came out to inspect their efforts and congratulated all of them on their excellent work, and furthermore suggested that they all retire to the big house for some well-earned tea. In a fairly large room, a sort of communal staff dining hall, the men sat drinking and talking. Self-consciously, Jack and his colleagues entered and received what they thought were some queer stares from the 'natives'. They were looking very windswept and dishevelled but were quickly made to feel at home. After showing all and sundry their cuts and splinters from the mast erection, they were chided by some of the residents and welcomed into the clan.

They soon learned that some of them, like themselves, had been voluntarily enrolled from trade and industry to assist in building small electronic units or other devices, as they were required, or even if needs be helping the fitters and riggers erect structures. They assured them that there was always something to do and judging by the spontaneous, enthusiastic spirit of these men, they knew they would all enjoy any task, however menial.

This not very technical and most un-secret work was thus Jack's introduction to Bawdsey Experimental Station, the first of all the radar stations, supervised by Robert Watson-Watt.

To save money, Jack usually cycled from London and camped near Bawdsey. This enabled him to enjoy a weekend of camping and at the same time to do some useful work at the station. At other times he enjoyed the luxury of a bed and breakfast in Felixstowe. He found the train trip out from London a trifle nauseating, and, being a keen cyclist, preferred the trip over the fairly flat countryside far more enjoyable than the stuffy train. Not that he criticised the services offered in those days by the London and Northeastern Railways. In fact, one enjoyable aspect was the excellent fruit lunch he could buy from the automatic vending machine, which in those days stood at the foot of the stairs at Liverpool Street station and which he could devour on route in luxurious, if sooty, comfort.

During one of his weekend visits to the station, Jack was dining in the main hall on a meal of sausage and mash. This typical English lunch had apparently caused the visiting boss Watson-Watt much amusement. He made an impromptu and delightful speech to those present who cared to listen, proving beyond doubt that the true Scots haggis was in every way infinitely superior to these little English sausages which in any case were principally composed of bread!

His command of English was such that it was a pleasure to listen to him speak, and his lilting Aberdonian made it quite easy to tell that he emanated from that northern part of the British Isles.

A few years later, Jack was to spend considerable time working on radar stations in that part of northeast Scotland, from which Robert Watson-Watt had originated. His pleasant personality and happy disposition were imparted to all his 'slaves', and his ideas on community discussion, all having the strong flavour of a socialist discussion group, worked magnificently. He had the capacity to make highly intellectual scientists work under almost intolerable conditions and achieve the most amazing results. Together with a few helpers, namely Arnold Frederick Wilkins, William Alan Stewart Butement, Edward George 'Taffy' Bowen, Donald Harold Preist and James Fleming Herd, he fathered the ideas that one at a time were brought to fruition.

The main radar was the CH, the Chain Home, which consisted of four 360-foot steel transmitter towers that held the aerial array slung between them, and three 240-foot wooden receiver towers, which supported the receiver aerial array. Probably because of his research in the 1920s into detecting the radio waves produced by thunderstorms, to track their position and movement, Robert Watson-Watt used 10-15-

metre wavelengths in developing chain home radar. His early work was given the name Radio Detection and Finding, which later became RDF, the acronym that was used for radar.

In addition to the CH there was the CHL (Chain Home Low), with its 40-foot-wide by 20-foot-high aerials that used 1.5-metre wavelength. Two other specialised radars, the ASV (Air to Surface Vessel), and AI (Aircraft Interception) also used 1.5-metre wavelength.

There was also the IFF (Identification Friend or Foe), developed by Don Preist, in which the components inside the small box in a friendly aircraft would recognise the radar signal from a radar station, amplify it, and send it back so that the blip on the cathode ray tube would appear brighter than those from an aircraft not using an IFF, which would tell the operator that this was a friendly aircraft.

The last of these brainchilds was the CRDF (Cathode-Ray Direction Finding) system, which was largely instrumental in winning the U-boat war and helped in a way that no other system could in the detection of the V2s at the very end of the war. It was initially developed by Robert Watson-Watt in 1926 for locating lightning and therefore thunderstorms, which were proving a menace for aircraft.

Edward Bowen and his assistants had difficulty in trying to comply with the limits on weight, size and power requirements of the AI radar. In his book *Radar Days*, Bowen recalls that, as if by a miracle, a small-tuned radio-frequency (TRF) receiver designed by the EMI company, for their projected television service from Alexandra Palace, found its way to Bawdsey from the EMI factory. It was about 3 inches wide by 15 or 18 inches long, had seven or eight valves, and operated on a frequency of 45 megacycles per second or a wavelength of 6.7 metres, with a bandwidth of 1 megacycle per second, which met their needs. He added that he suspected it had come 'through the back door' and not through official channels.

Working at EMI and assembling, installing and maintaining television sets in central London, Jack would have been familiar with this receiver, and his skills must have been a real boon to the scientists working at Bawdsey.

The main capabilities of the CH (Chain Home) radar were long-range performance coupled with the important ability to read the height of an aircraft within 1,000 feet. Though its height and range information was fairly accurate enough for day interceptions, its bearing accuracy at times left much to be desired. It had directional ambiguity, which sometimes allowed the operators to read an aircraft as being in front of the station when in fact it was behind. This rather unfortunate idiosyncrasy led to the sad battle of Barking Creek, when

RAF fighters shot down British aircraft over the Thames estuary. This not too enticing introduction of the radio location system occurred a short time after the outbreak of the war.

This ill-fated episode occurred at the time when King George VI was on a conducted tour of Fighter Command Headquarters at Stanmore. The Air Officer Commanding, Hugh Dowding, was explaining to the king the working of the filter room, when an aircraft not showing IFF identification friend or foe, the radar transponder identification signal, was plotted by a Chain Home station. Being on the Thames estuary, the mistaken bearing showed the aircraft to be 30 miles out to sea towards Germany instead of 30 miles behind the station and safely inland. As no IFF signal was shown on the tubes of the radar, the aircraft supposedly over the North Sea, had by the rules of the game to be regarded as possibly hostile and was plotted on the Stanmore table as X-raid. Fighters were scrambled to intercept and take the necessary action. These fighters speeding out to sea were plotted by another radar station and also not showing IFF, appeared on the table at Stanmore as another X-raid and further fighters were scrambled to intercept them. The confusion resulted in the aforementioned Battle of Barking Creek, which unfortunately ended in death and destruction for a number of the unfortunate participants. The whole sad incident led to a tightening up of flying discipline and the importance was stressed to the pilots of switching on the little IFF switch.

The small box holding the fighter's identification radar was mounted behind the pilot and it was not policy to lecture all and sundry on the reason for this little gadget. It is hard to understand now, but the pilot of that time, with no knowledge of the radio location system, was quite unable to appreciate the importance of switching on his IFF set. Sometimes, the keen pilot inadvertently flicked on the toggle beside the IFF switch and in this way blew the set to smithereens. This second switch ignited a detonator enclosed in the set and was only meant to be used for its destruction in the event of possible capture by the enemy. However, the thud followed by a shocking smell in the cockpit did not deter many pilots from carrying out this damaging operation time and time again. Eventually, the destruction switch was secured with a thin piece of wire to prevent its accidental use. It goes without saying that after destroying his IFF set, the fighter pilot would henceforth be plotted by all radar stations as a potentially hostile aircraft.

There were many teething problems with the radio location system, but the Battle of Barking Creek was one of the saddest. This CH directional ambiguity was to make life for Air Vice Marshal Keith Park very difficult once the Battle of Britain started. Ghost tracks of aircraft led to much wastage of engine hours and more importantly the wild goose chases caused the pilots much exasperation and frustration. Nevertheless, once the Battle of Britain started, the CH stations as the backbone of the radar system bore the brunt of most of the air fighting.

Another group at Bawdsey was concerned with the attempt to devise an airborne radar capable of plotting shipping. This was coupled with the requirement to produce another airborne radar to detect bombers at night and enable the night fighter to home in on them. For a time, a comparatively long wavelength was tried in an effort to make the airborne radar work, but real success in this field of endeavour was only followed when a change was made to the very short wavelength of that time, 1.5 metres. This form of equipment, bearing the initials ASV (air to surface vessel), was used successfully throughout the rest of the war and gave rise to various options, which are in use to this very day.

Doctor Edward Bowen carried out the development of ASV and its sister radar, the night fighter AI (aircraft interception) radar to a successful conclusion.

Under the stringent war service conditions, these items of equipment, which came to fruition only in the nick of time, met with fantastic success.

At the same time at Bawdsey, in an army development site, Dr Butement was developing a radar station, the CHL which could detect surfaced U-boats and small surface vessels. He was preparing a radar gun-sight to enable land-based guns to open fire on enemy shipping picked up in the radar sight. He also chose the very short wavelength of 1.5 metres.

By splitting the narrow beam and balancing the signals from the left and right beams, he managed to obtain a considerable degree of bearing accuracy. The story has it that when Winston Churchill came to the site for a demonstration, a small vessel was picked up by the radar and the aerials of the radar were kept focused on the ship. 'Naval Person' Churchill, on reading the range of the vessel, insisted that if it were at that range, under these conditions of visibility, he should be able to see it visually. He squinted up a telescope with a hair sight mounted on the bearing indicator of the aerial to look for the ship, which should have been directly ahead. He puffed and puffed on his cigar, but no ship could be seen. After a few seconds had passed, everybody breathed a sigh of relief when the ship, viewed end on, emerged from behind the hairline.

Though this story sounds a bit far-fetched, Jack worked on this kind of CDU/CHL equipment when it was further developed and found a high degree of accuracy. Usually, however, the electric motor rotating the switch contacts made the trace jitter so much that it was at times quite difficult to even see a small vessel.

These systems, the CH, CHL, AI and the ASV, developed before the war, each came into service just in time to close the yawning gaps in Britain's defences.

It is acknowledged that at the outbreak of the Second World War, the German radar systems were more advanced than the British. These included Freya, a rotating air search radar used 1.2-metre wavelength; Seetakt, a naval surface search and gunnery radar using 80-centimetre wavelength; and Würzburg, a highly accurate search and fire control radar using 53-centimetre wavelength, which was considered very advanced for its time.

The British had Chain Home (CH) that operated on long wavelengths of 10-15 metres, providing long-range early warning but with limited accuracy and fixed installations. Chain Home Low (CHL), meanwhile, was similar to the German Freya, using 1.5-metre wavelengths for lower altitude detection.

However, Britain integrated its radars into a comprehensive system devised by Air Chief Hugh Dowding, while Germany's shorter-wavelength radars, though technically good, were not part of such an integrated network.

The Germans did not continue developing radar in the same way the Allies did. Hitler and some of the Nazi leadership did not fully appreciate the strategic value of radar. Hitler was more fascinated by the V1 and V2 rockets, jet aircraft, and massive tanks rather than systems like radar. As a result, resources, funding, and engineering talent were diverted from radar research toward these high-profile projects. Though there is no written evidence to confirm this, in 1943, Hitler reportedly dismissed radar's importance, saying 'We have conquered Europe without radar, we can win the war without it.'

In September 1939, war was declared and as Bawdsey, which was on the Suffolk coast, was considered too vulnerable, its staff and scientists were relocated first to Dundee, which was too small to accommodate all the people involved, then to Worth Matravers, and then, because that was too near the channel coast, to Malvern. Jack reluctantly had to bid everyone goodbye, as he could not move with them. They took on the name of Telecommunications Research Establishment (TRE).

Jack's brother, Harold, was born on 12 February 1939, his maternal grandfather died on 19 October 1939, his father died on 13 November 1939, and his maternal grandmother in May 1940. With his father's death, his mother was declared to be an enemy alien, as Aaron had not obtained naturalisation. She and Harold were evacuated to Stalybridge, but after Aaron's brother and Marie had used her birth certificate to prove to the authorities that she was British, she was allowed to return to London and a flat was found for her in Mattock Lane, Ealing.

CHAPTER 3
THE ROYAL AIR FORCE

In early 1940, Jack decided to volunteer to join the RAF as he wanted to fly. He paid a visit to the balloon centre at Stanmore, not because he had the slightest interest in balloons, but because Stanmore was, of course, Lord Dowding's Fighter Command headquarters. This was the one and only filter (or the receiving centre) for all information on aircraft as plotted by the radar stations. Backed up by audible and visual reports issued by the Royal Observer Corps, the presiding command controllers could immediately identify all aircraft plots as friendly or hostile. Security had been completely tightened. After going for a short interview with a rather junior officer, Jack was told to go home and sit tight as he would soon be called up in the new special organisation to be set up for RDF men.

Time went by and Jack continued his civilian radio work in the West End of London, anxiously awaiting his call up, but maddeningly, nothing transpired. His attempts to visit the radar stations with which he was acquainted were now squashed and in desperation he went to Uxbridge RAF station, just to the west of London, and made discreet inquiries about his future. Here he was immediately inducted into the new RAF branch. Eventually, he learned they were to be called 'Radio Mechanics' to distinguish them from the ordinary wireless electrical mechanics of the RAF.

Eventually, after a short period of RAF disciplinary training at Uxbridge, Jack was sent off to Yatesbury, in Wiltshire, to the Number 1 Radio School of the RAF. On arrival at this the original wireless school of the Royal Flying Corps, he found that they were strange fowl amongst the dozens of would-be wireless operators, as nobody seemed to have any idea of what they were supposed to do.

He also met one or two of the men with whom he had worked in the past and together they were detailed to assist in the building of the first radar school at a spot some distance from the main station.

This time was spent in creating new secret equipment for the proposed establishment and it was while engaged in this task that Jack had an extremely depressing experience while un-crating a top-secret radar receiver known as the RF6, manufactured by the Cossor Wireless factory. This device was the acme of radio receivers at the time and incorporated every known device to secure superb results. It was an immensely secret machine, and each and every one of its dozens of features likewise. It was a triumph of receiver design and demonstrated just how far forward the science had already progressed on the drawing board, as well as in the factory.

As this unit, which was brought down under special guard from the Cossor factory at Highbury in London, was carefully unpacked, the men held their breath in awe as they gazed up at the 8-foot-high monster. Carefully uncovering its rugged face, they were stunned to see scrawled across its beautiful black face in white chalk an enormous hammer and sickle. The political humourist who had presumably completed his artistic effort during the crating stage at the factory did not even have the decency to avoid the delicate meter faces, the glasses of which had been pushed in and the internal movements damaged. This deliberate vandalism by their own people depressed everyone.

On some weekends during this period of the phoney war, Jack and his colleagues made the 90-mile dash to London to sample the fleshpots of entertainment in the West End. During one of these jaunts, Jack went to the London Palladium and saw his favourite team of comedians, The Crazy Gang, Flanagan and Allen, Nervo and Knox, and Naughton and Gold. Even in this hilarious show, *The Little Dog Laughed,* they did not allow the audience to forget the war. Flanagan and Allen sang the popular tune of the moment 'Run Rabbit Run', a song that originated from one of the odd spots of German air activity over the Shetland Islands. The newspapers had reported that a German aircraft had machine gunned a harmless little rabbit. Hence the title. In fact, there was more to it than that. A German aircraft had carried out a reconnaissance of a new wireless station being constructed in the Shetland Islands, way north of the northernmost tip of Scotland. The station was CH Netherbutton, the most northerly of the CH radar stations of the Home Chain.

While Jack was at Yatesbury, he received a short visit by three French air force officers. They had been given official permission to be shown around and observe Britain's radar activities. In their elegant, dark-blue uniforms, they tip-toed around to avoid contaminating their shoes with English mud, talked to nobody, and in general gave everyone a pain in the neck. In an effort to establish a local entente cordial with Britain's

Allies, Jack tried to converse with the men in his schoolboy French. When an urn of tea was sent out from the main camp, he attempted to get beneath their aloofness and apparent disdain, which at first he had attributed to their inability to speak English. Diplomatically, and struggling with his strained vocabulary, he opened the conversation by criticising the quality of the NAAFI tea, the expressions of contempt on their faces doing little to assist his attempts at polite discourse. After a few words he discovered that they could speak excellent English, certainly a lot better than he could speak French, but as ever, their handsome faces wore an almost perpetual sneer, the reason for which soon became apparent. 'You Tommies don't expect to beat the Bosch with bits of wood and wire? Surely you know they have thousands of armoured tanks and bombers. What good will all this be? Why indeed were we sent here to waste our time in this cold, windy and muddy place where it never stops raining?'

Jack responded by pointing out that when the German armies eventually attacked and had broken their heads on the Maginot Line, they might try to flatten France with their bombers. His job was to prepare, as quickly as possible, our own and Allied territories against air attack and this they were doing to the best of their ability. He told them that some units had already arrived in France; one under the famous technical writer and radio expert Flight Lieutenant John Scott-Taggart, who was already established there.

The youngest of the Frenchman told Jack that many French officers had already established a secret pact with their right-wing organisation, which still enjoyed a form of contact with the Nazis. This group, he said, was well supported by the officer class of the three French fighting services. They could see no point in starting to fight and felt that negotiations with the Germans, before the main onslaught started, was far preferable to immediate defeat by the Luftwaffe and the German army.

Jack was staggered by the flagrant defeatism of these French airmen and realised the seriousness of their statements and the possible existence of a major security leak. He brought the information as it was explained to him to the attention of one of the RAF officers on the site, who laughed it off as mere talk. The officer, being a rather keen radio ham, regarded the whole war as a wonderful game and an excellent chance to get to grips with the new and fantastic radio devices he could never afford in peacetime.

Jack and a few of his colleagues were at an RAF signals camp just after Dunkirk, in May 1940. Under secret cover, a request had been made for some temporary, technical RDF assistance, the exact duties of which were not specified. There was still a minute number of people with radar training and in his opinion, it was an honour to be able to join in the air battle that was obviously getting underway. The Luftwaffe was now making probing attacks around most of the English coastline.

Collecting their small kit and enough clothing for about a week, they climbed aboard an RAF truck and headed for Swindon, where they entrained and travelled east to London. There were six in their party and they had plenty of time during the journey to discuss what fate might possibly have in store for them. The phoney war had ended, and the German army had passed through France like a knife through soft butter. It was good to know that they would be doing something constructive towards the war effort.

They were routed through Bromley, a London suburb, where 60 Group had its local Wing headquarters. Here they were broken into pairs and John Strong and Jack climbed onto the back of a small Bedford truck. They made towards Maidstone and Canterbury, a road on which, in happier days, Jack had travelled many times on his racing bicycle. As they approached the narrow winding streets of Canterbury, the air was heavily pockmarked with the bursts from ack-ack shells from the German aircraft who were now beginning to appear in quite large numbers over the Kent countryside. They stopped for a short time in the narrow lane and watched the intense activity overhead. Though there were very few of their own fighters, they were giving a very good account of themselves. The jarring rattle of the guns and the horror and death stalking the clouds over their heads was completely out of keeping with the beautiful green countryside. Eventually they carried on and made their way down to the coast at Broadstairs, a pleasant seaside resort on the North Foreland.

There was plenty of air activity and Jack and his colleagues accepted this without any particular interest and made no attempt to define details of any particular operations. They had become quite blasé to the whole business. From Broadstairs they were routed to the northernmost tip of the North Foreland. They dropped off the back of the truck at the entrance to one of the new CHL stations that were already coming into service and acting as low cover for the CH stations that could not see aircraft much below 3,000 feet.

After their identity cards had been carefully checked at the makeshift orderly room, they were placed on duty immediately. They had no idea of exactly what to do, but common sense readily made-up for their lack of knowledge. This, one of the ramshackle almost homemade stations that Dr Cockcroft himself had recently commissioned, had already done great work in combating the magnetic mines at the approaches to the Port of London. From Foreness, looking northeast one could scan the Dutch and German skies. Due west they could look along the Kent coast towards London and the Thames estuary. Looking due east, they enjoyed a long-range snoop of German air operations in their newly acquired French aerodromes. Although the Germans usually carried out their air movements under complete radio silence in an effort to prevent the British developing an intelligence picture, they did not realise that the British were meanwhile plotting their build-up of formations in French territory long before they set course for England.

Although the 20-foot-high CHL station masts were dwarfed in size by the 360-foot masts of a CH station, the small unit was a giant in technical performance. Jack and his colleagues were there to assist and do anything that was required; to act as operators, or if the equipment broke down to assist in its repair and to be on standby in case of breakdowns. On the short strip of the Kentish Foreland coast there were two other giant radar stations that were already doing trojan work. The first was Dover CH and further west CH Dungeness. Although the CHL could not plot at the speed of a CH station, they had a far greater accuracy, and their plotting was not marred by the serious ambiguities that often accompanied many CH plots of enemy aircraft and formations. The CHL, however, could not read the height of an aircraft or the formation like the huge CH station.

At that time there was a great scarcity of technical men available for the operation of this radically new type of radar equipment, and the station was shockingly short staffed. Due to the highly specialised nature and secrecy of the work, nobody could be brought in from outside to assist and long hours were the order of the day and night. Like all other radar units, the station was operational 24 hours a day, which included a supposed hour a day off the air for maintenance. During this hour, the local CH station, Dover, was always warned that the CHL station would no longer be giving low cover and should therefore be doubly alert.

In the small operations room of the CHL, the technicians rotated the hand pedals which turned the roof aerials, while their eyes were glued to the green flickering 3241 tube waiting for the appearance of an aircraft blip. The drills were quite simple. The range of the aircraft was read off the screen and the bearing read off the crude, but accurate, scale mounted on the aerial drive column. The resulting plot of range and bearing was marked with a pencil on the tracing paper clamped over a military grid map and the four-figure coordinate pinpoint read over the telephone to Fighter Command HQ in Stanmore.

It was a most exciting station as German aircraft cruising over France could be seen at almost any time of day or night. German aircraft flying at 10,000-15,000 feet could be clearly seen up to 75 miles inside France, where the Luftwaffe was now taking up residence.

After a thrilling spell in the operations room seeing and plotting aircraft, it was time to take a turn in the transmitter hut; once again this entailed turning by hand the pedals that rotated the massive transmitter aerial, but this time blindly following the receiver aerial by the simple expedient of keeping the needle of a small electrical meter continuously on zero. In the early hours of the morning, this could be an extreme form of torture.

Sitting in a small hut with a transmitter towering over you and droning its way through the night, was in itself no hardship. Though these early transmitters used the enormously high voltage of 25,000 volts, that was not in any way dangerous as the thick sheet steel sides and excellent manufacturing technique made the transmitter perfectly safe from X-ray radiation.

The T3056 had one rather hair raising, if not dangerous idiosyncrasy. At times, usually during the early hours of the morning, when the air became damp, the high-voltage system broke down the air insulation and a fantastic scream ensued from the iron walls of the transmitter. The screaming 25,000-volt cacophony vibrated the heavy steel walls of the transmitter as if they were made of rice paper. This ear-splitting racket occurred with no warning and invariably in the early hours when the men were battling to keep their eyes open and the needle on zero.

Outside the transmitter hut, everything would be peaceful and quiet with the Luftwaffe across the channel in France all nicely tucked up in bed. Jack and the others, however, were on a twenty-four-hour watch and dared not relax even during the quiet and peaceful predawn hours. The screaming corona would suddenly burst upon everyone, all but throwing them out of their chairs in fright. As they left the pedals to stop the racket, the telephone would invariably scream hysterically and the operator would demand that the aerials be put back in line. On hearing the corona racket over the telephone, the operator would quickly understand the situation and sympathetically hang up.

The standard treatment was to wind the transmitter high voltage system down 2,000 or 3,000 volts and then once the corona had stopped, to gradually increase the voltage back to its normal 25,000 volts. Needless to say, the acoustic shock left everyone with palpitations for some ten minutes afterwards.

During daylight hours there was now intense activity, both friendly and hostile, although there seemed no great danger to CHL Foreness. Having been allocated a BSA motorcycle in case urgent spares were required, Jack and other RDF personnel often made trips to the CH station at Dover and even further afield.

A friendship was struck up with the daily changing occupants, pilots and ground crew, of the fighter aerodrome dispersal tent about 2 miles inland. Sometimes they kicked a ball around, while chocolates and sweets were always brought and exchanged.

On one occasion after a short and hectic game of football one afternoon, Jack and his companion lay on the soft grass panting and the pilots returned to their bell tent for a rest. All was quiet and peaceful. They vaguely heard the muted note of some aircraft engines, which John, with his excellent eyesight, located while they were way out to sea and coming in their direction. Low flying aircraft from the three aerodromes in their vicinity, Hawkinge, Lympne and Manston, were always around so the appearance of two aircraft cruising low over the sea seemed to be just another patrolling section of fighter aircraft.

As there had been no fighters scrambled, the men paid little attention to them as they grew larger, and in no time they were almost overhead. In the blinding sunlight they stood up to greet them as they cruised low directly over their heads, when with a shock they saw the huge black crosses of Messerschmitt 109s. The German pilots must have seen them but decided they had more important things to do with their ammunition than to strafe a pair of stupid civilians. Jack and John were thankful for the fact that they flew right over the blast hill and camouflaged fighters without seeing a thing and flew serenely inland. The two or three Hurricanes stationed there could not be seen from the sea as they were located snugly behind a mound that acted as an anti-blast hill masking their presence. Camouflage netting, plus the natural camouflage paint on the Hurricanes, made them extremely difficult to spot from the air.

Running to the dispersal tent, Jack frantically wound the field telephone and reported the two enemy aircraft, which nobody had seen or reported. Flying very low, they had avoided, intentionally or otherwise, detection by Allied radar.

It was apparent at times that the Luftwaffe now regarded this forward area of Kent as part of their own happy hunting grounds.

On another occasion, great panic reigned in the technical division of the station when the one and only oscillograph test unit gave up the ghost. Being a Londoner and knowing his way around the Cossor factory, Jack was delegated the task of taking this invaluable test instrument to London for speedy repair.

With no rest and after a night's duty, he climbed aboard a 500cc BSA motorcycle and made for the east coastal road. For purely personal reasons, he preferred this route to the trip through the narrow streets of Canterbury. From Margate, the smooth surface of the road and steady hum of the engine gradually helped him relax. Roaring along the leafy Kent lanes, he reached Chatham and commenced the last leg of the trip to London. Without noticing, he soon realised he had driven straight through a red traffic light. The guilty feeling in the pit of his stomach gradually receded as he waited for the strident gong of the ever-watchful police car. Strangely enough, however, it did not come. They were too busy, he supposed, watching the fires and the expanding ceiling of black smoke spreading over the London skies.

Jack was extremely tired as apart from a few hours on the transmitter floor, he had not slept properly for days, and it was becoming increasingly difficult for him to keep the heavy motorcycle heading in the right direction. The glaring red ball of the setting sun was now right in his eyes as he drove along the Rochester Way towards London. With a belated and frightening wrench, he brought the machine back on to the road. That was a near thing!

He could not afford to lose the test oscillograph at this stage in the game and felt behind to see if the instrument was still secure on the pillion. The shock of the near accident had woken him up. Nearer London than he thought, now Blackheath Hill, and the road had changed direction. The enormous frightening black clouds from the burning oil storage tanks across the Thames estuary at Tilbury now dimmed the blazing disc of the sun that seemed to have burned a hole in his tired brain.

It seemed that the Luftwaffe could do as it pleased, burning and bombing wherever it chose. The stacked formations of bombers at times seemed to swarm like locusts and fill the sky. Worst of all, they did not seem to be able to do anything to resist the masses of squat, black German bombers.

True, he had seen one or two crumpled Heinkels lying in fields near the east coastal road somewhere near Herne Bay, and also a burnt-out Messerschmitt 109 in some tennis courts at Chatham, but what were three aircraft compared to a sky blackened with a thundering great horde of bombers? Strange how those people were still playing tennis doubles within a few yards of the crashed Me 109. Some people did not seem to know there was a war on!

On the face of it there could only be one end to it all. Britain simply did not stand a chance. If only they had listened to Churchill a few years ago, all this horror would never have happened. Now, with all the petrol going up in smoke, even the few fighters might be grounded.

Jack drove through the vibrant echoing tube of Blackwall Tunnel under the Thames, before heading through the city and on to Highbury and the Cossor factory. 'I hope they can fix the 'scope quickly,' he thought. It needed a new mains transformer. Could they fit one without too much form filling? RDF sets could not be maintained without an oscillograph. He was halfway across the city before he vaguely realised that the Cossor factory might in any case be closed. The RAF was working 24 hours a day, but the factory workers were most likely already at home; maybe even watching the rising angry black clouds from the Tilbury fires.

Never in his wildest dreams had Jack thought this would ever happen to London. He pulled in near a public telephone booth. His legs were weak and it was all he could do to balance the bulky motorcycle with the weighty oscillograph mounted on its pillion against the high kerb.

The telephone rang and rang, but there was no answer from Highbury Grove. Now what? Sleep? He imagined clean fresh sheets into which he could sink into luxurious oblivion. He needed to sleep somewhere. Climbing back onto the saddle, he headed west along High Holborn.

If he had not been so tired, he would have stopped to look at the shop windows, where in happier days he had bought radio parts. Could that have been only a few short months ago? Apart from some air raid wardens and tall, helmeted policemen here and there, the roads and pavements were practically deserted. He circled round Marble Arch and on to Shepherds Bush, before the last lap to Ealing village, where he had family that would accommodate him for the night.

Making the BSA secure, he struggled with the oscillograph to his uncle's flat; the name on the bell push seemed so neat and civilised, but there was no answer. He sank dejectedly on to the sturdy steel case of the test instrument, when suddenly, like a Messiah, the caretaker arrived and opened the flat for him. He staggered through the door with the weighty valuable instrument in his arms and gingerly placed it on the carpet in the corner, out of the way. He needed a bed badly. As he entered his aunt's bedroom, a photograph caught his eye. It was of his two uncles taken in the First World War. Two young, proud, confident faces smiled out of a drab brown photograph surrounded by the coloured flags of all nations. At least they had won their war.

This war seemed lost from the word go. From what he had seen, the British did not stand a chance. Without removing his clothes, Jack lay exhausted on the bed in morbid thought with his eyes closed.

'Wakey wakey! Come on, Jack. Wake up. Dinner's ready. Get yourself cleaned up.' His mother and aunt were standing in the room, and it took him some time to realise where he was.

So that was what life had come to. His uncle was helping to fight one of the most dangerous fires in England's history. His little brother, instead of playing hopscotch, was trading bomb shrapnel and those people in Kent had the audacity to play tennis with a burnt out Messerschmidt in the end court. There was a moral there somewhere, but he was too tired to see it at the time. He thought he had better get organised now as he did not want miss Winston Churchill's broadcast.

Early next morning, despite having stayed up most of the night listening to his uncle's joyful description of his activities in the great fire, (sure he enjoyed every moment), Jack headed to the Cossor factory with the oscillograph, and later that day and after receiving VIP treatment at Highbury Grove, he hot footed it back to CHL Foreness and the war.

After returning to Yatesbury and some more training, in late 1940 he was sent to Rosehearty in Scotland. CHL Rosehearty was on the clifftop and was an extremely windy and freezing-cold site, but perfect for spotting enemy planes flying in from Norway.

It seems that the coming of spring was no safeguard against the snow in that windy northeast corner of Scotland. It was already snowing when Jack and his fellow RDF personnel went on duty one night for the evening watch and sat shivering in the unheated huts of the CHL station, which needed all the electric power that could be supplied by the 15 KVA diesel electric generator. There was no reserve power for electric heaters and at this time no paraffin stoves were available.

Housed in the same compound as the CHL station were the men of the Gordon Highlanders, who provided the security guards. They paraded around the perimeter, stamping their feet and blowing the snow from their moustaches. They greeted Jack and his friends with the typical Aberdonian 'Fine day', which was the stock in trade greeting even if it was raining cats and dogs. As the snow fell from the grey skies, it grew dark, and a vicious northeast wind came up. The wind abated somewhat and then the snow began to come down in earnest.

It snowed throughout the night and by the morning the ground was covered in a mantle of beautiful, clean white snow. After the night watch they donned their greatcoats and trudged their way down to their billets in the village of Rosehearty. Feeling tired after the long night's duty, they staggered and plunged their way down the hill. A hearty breakfast of 'brose', a simple Scottish dish, traditionally made from oatmeal and hot water, sometimes with the addition of butter and salt, with sausage cooked by Mrs Buchan the landlady, by 9 o'clock Jack was fast asleep. It snowed steadily throughout the day with the thick flakes dropping vertically down. Due to the abject shortage of staff, they were to go on duty again that night. With his host Mr John Buchan, Jack inspected the air raid shelter which they had tried to cut out of the rocks. It was obliterated now by the heavy fall of snow. After tea, Jack pulled on his two pairs of sea stockings and fitted his gumboots over the top of them. With two jerseys under his RAF tunic, followed by his greatcoat, service gas mask and the sou'wester, he collected his friends of the watch and they ploughed their way very slowly up the hill to the radar station.

The wind had been increasing, and the snow was now at times assuming a horizontal trajectory as he staggered into the operations room. They arrived in the middle of an animated scene. The hand-rotated bicycle chain that turned the windmill-like aerial some 20 feet over their heads, had snapped in the gale. Two members of the duty crew were on the roof of the hut holding the massive aerial array, whilst an air ministry technician, Archie Blair, effected the repairs. With a flourish, the I/C watch signed off in the logbook and they took up their stations for the evening. One man went to the transmitter hut to bind the transmitter aerial, and the other three members of the crew settled down in the operations hut, about 50 yards away. Snow or no snow, the CHL had to keep on the air and survey the skies over the massive shipping convoys coming round the top of Scotland at Thurso and the headland at Fraserburgh and Peterhead.

Already the heavy snow had claimed some victims. One local man who had missed the road up from Macduff, a small township some 3 miles from Rosehearty, had been found by a policeman. He had died of exposure. Life for these hardy folk was indeed a trial during the winter months.

Stanmore informed Jack that there was very little happening. The telephone link with civilization made life bearable. It was strange to be exchanging chatter with people a short bus ride from his home in London while they sat 700 miles north watching for enemy raiders. It was impossible weather for aircraft to fly, but information on convoy movements on the heavily mined route was always needed by the Naval Liaison Officer at Aberdeen naval base 70 miles south.

Though the room was cold, with a blanket or two they achieved a measure of protection from the icy blast outside. Careful doctoring of the blackout shutters with blankets made the room comparatively snug, but certainly not warm. There were of course no beds, so those men not actually watching the screen or operating the transmitter were usually stretched out on the floor behind the receiver, where the warm air from the power plant circulated. The hard concrete floor was not exactly the most luxurious of beds, and men always woke for a turn at the tube with a sigh of relief and usually a very stiff neck from an errant draught.

This method of rest was most conducive to nightmares. One of the Air Ministry technicians, a short plump young Yorkshireman, Bill Iredale, with an accent that could be cut with a knife, let out a terrific yell one night in his sleep. In a panic, Jack had torn round to the back of the receiver to see if he had accidentally electrocuted himself by touching the 6,000-volt power supply for the cathode ray tube. Bill was kneeling, facing away from Jack, with his head down on the ground and his bottom up high. With his fingers he held onto a small duct on the floor enclosing the power cables. He was fast asleep in this strange posture and also in the middle of a nightmare. As Jack arrived, he yelled out at the top of his voice in his broad nasal Yorkshire accent 'Ah'm oopsahd dahn, ee, ah'm oopsahd dahn.' Jack woke him up and Bill opened his eyes in relief when he found he had just been dreaming. Apparently, he thought he was falling over a cliff and that his last moment had come. Bill, though dour, was a likeable and humorous character and Jack always found it a pleasure to work with him. Radar people generally did seem to be a cut above the average.

There was a Devonshire man, however, who was not so pleasant during their nightly sojourns. Although Jack described him as a wonderful chap with an excellent supply of jokes and an indestructible sense of humour, Ray Thorne, or 'Dartmoor' as he was affectionately known, had a very antisocial habit. Despairing of the local Scots beer, he had arranged for the dispatch from his native Devon of a barrel of rough Devonshire cider and insisted on bringing a quart or two of this thick, green, oily liquid on watch with him. At midnight his snack usually consisted of a huge chunk of cheddar cheese, some pickled onions and bread, which he swirled down with this corrosive liquid from the south. As he chewed, his eyes swivelled around in his thin face, watching the others with devilish glee. In a short time, he would have the operations room to himself. It was just as well that he could not rotate the aerial and read off the plots to Stanmore without any assistance, as he made the atmosphere simply unbearable for the rest of the men.

Outside, the gale was piling snow up in enormous drifts and the buffeting wind made it impossible to turn the aerial. The struggle with the binding pedals was terminated by the telephone ringing from the transmitter hut. The worst had happened; the bicycle chain had snapped and had snaked upwards to be ground into the gear wheels. It was irreparable and the 20-foot-high by 40-foot-wide aerial was now in danger of being smashed by the gale-force wind that was hammering the aerial system backwards and forwards. Jack put on his greatcoat and sou'wester and told an operator to report the station off air, before heading out into the wind and snow to the transmitter hut with some thick rope to lash the aerial in an easterly direction pointing towards Norway.

Jack climbed the 20-foot ladder and hung onto the wooden frame at the base of the aerial system. He crawled around the turntable to the other side of the gantry, while another operator came up to steady the aerial from the ladder side. While he was preparing the lashing, a vicious gust caught the aerial and knocked his companion from his perch on the ladder onto the ground. Another operator took his place, and they lashed the aerial array facing the usual direction from which hostile aircraft came, the east. Luckily, Jack's fallen friend had landed on the roof of the hut in the soft snow, unhurt.

Looking over towards the barrack hut that housed the Gordons, which he could now only vaguely see, Jack was surprised to note that it had almost completely disappeared under a snowdrift. On the windward side the snow had already reached the top of the blast walls of both huts. Together Jack and his companion staggered back to the operations hut, leaning almost horizontally into the wind and snow, and took off their sou'westers, which now stood up by themselves on the floor. They were stiff with ice.

This was not the end of their troubles, as they now found that the telephone line to Stanmore was no longer working. They decided that to avoid damage to the now useless receiver aerial, they had better lash it on the same bearing as the transmitter aerial. It was now intensely cold and Jack phoned the Gordon guards to see if they could give them some food. Fortified by some bully beef and biscuits and some black water cocoa, the men lay down together to try to get warm with the aid of a blanket or two. Jack sat down in front of the radar tube and watched, with a telephone handset in position just in case Stanmore came back online.

Slowly the night went by and by 8 o'clock the next morning, the blizzard had not abated, and Jack began to worry about the diesel fuel position. Putting on their greatcoats and sou'westers once again, they fought their way to the transmitter hut besides which the diesel storage tank was located. Even the transmitter hut was now very difficult to find. They had to dig their way down through the snow to refuel the diesel-powered generator that chugged steadily and comfortingly on. If it had stopped and cooled down, they would never have been able to start it again.

Jack checked that the man in the transmitter hut was comfortable and as he opened the door, he was greeted by a most satisfying smell of fried kipper. They were too late, however, as their crafty friend had already disposed of the delicacy. Apparently, he had cooked the kipper by switching off the transmitter, fixing the interlock on the 25,000-volt V1901 rectifier valve, and hanging the kipper around its neck. This valve operated at an enormously high temperature and cooked the kipper to a turn! Trust an Irishman to think up such an artful aid to cooking. This incidentally explained the fishy smell Jack had so often noticed when entering the transmitter hut.

The Irishman was happy, so they left him and this time they struggled madly to find their own operations hut. It was almost impossible to breathe in the blizzard, even behind their greatcoat lapels. By now it was long past breakfast time, and they were all hungry, but there was no way of even getting to the Gordons, despite their close proximity. With telephones out of action they were cut off from the world. They stood their sou'westers where the water would drain away and decided to wait until the blizzard had subsided.

By that evening they were still waiting, and, to make matters worse, one of the older members of the crew was having stomach cramps, mainly, Jack supposed, from the cold and lack of food. Food had never really worried Jack greatly and he was feeling fine despite the cold. After a particularly bad spasm from their elderly colleague, Jack decided to try to reach the Gordons' hut in an effort to obtain some cocoa. This time he fought to open the door and was appalled at the cold blast; it was quite impossible to breathe outside, and he struggled to shut the door as quickly as possible.

He was, however, determined to give it a try and in his full regalia, but this time with two sou'westers he started out with a gas mask in position to allow him to breathe and a long five-cell torch for light. It should still have been light, but the blizzard had wiped out any vestige of daylight. He stood for a moment outside the blast wall of the hut and decided in which direction the Gordons' hut must lay. He started walking, his gumboots sinking deeply into the snow, and walked as well as he could with the snow constantly filling the glasses on his gas mask. He switched on the powerful torch and for a moment he thought it was not working. Putting its glass flush with the glass of his gas mask, he perceived a faint glimmer. Turning around, he noted with dismay that the holes made by his gumboots were rapidly filling with snow.

This was his only means of finding his way back to the operations hut. He decided to go back immediately, and, as he retraced his steps, with the footprints getting fainter and fainter, he panicked and plodded and half fell back into the remains of his footsteps until they were blotted out. He carried on, intending to persevere in the same direction, and crashed face first into the blast wall, which he could not see. Feeling his way around, he eventually found the opening and then went through the door, stumbling into the middle of the room where he leaned on the receiver for support, trying to regain sufficient strength to remove his various wrappings.

Some time later, he tried to make the 50-yard trip again, but with the same results. It was a hazardous undertaking, because if he did not find his way back to the operations hut, he would have suffered the same fate as the man from Macduff. It was impossible to see and difficult to breathe out there in the blizzard, and without shelter nobody could have survived.

In an effort to raise both the morale and temperatures of the men, Jack started a sing-song and got everyone to tell jokes. Their enthusiasm, however, had already flagged, as the prospect of spending another night in the hut with the blizzard howling outside appealed to no one. Taking the 4-foot sheet-iron side covers off the radar receiver, he made the invalid sit between them so that his body temperature might be maintained by the hot air emanating from the hundred or more valves in the receiver.

An uncomfortable night passed and when he judged it must be daylight, Jack went to the door and noted that the steady gale had dropped, and he was pleased to see that the snow was now falling almost vertically. Only now and again did a gust stir the peaceful snow and remind them how violent the blizzard of the night before had been. The sky seemed have become a fraction lighter. Putting on his greatcoat and sou'wester, he once again attempted to locate the Gordons. Before he departed, his invalid friend pleaded with him to try to get some water. They had tried warming snow, but this had simply made him retch.

Through the steadily falling snow, Jack walked in the correct direction and when he reasoned that he had long passed the hut, he still could not find it. It was too ridiculous for words, but the heavy curtain of snow still marred all visibility.

Without much difficulty he managed to return to the operations room and a bit later tried again, but again without success. It was absurd, as the distance involved was so short that he simply had to strike the long barrack hut that housed the Gordons.

Once again, Jack walked out and this time he was determined to succeed. He was having no difficulty in finding his way back, so he resolved to stay out until he had located it. He ploughed through the snow, plodding steadily forwards. He had to be careful as he had no desire to make contact with the barbed wire that enclosed the minefield that surrounded the camp.

After a while Jack knew he had passed the hut. He looked up, the sky seeming a bit brighter, when suddenly, without any warning, he found himself falling through space and with his heart in his mouth, he sank deeply down into the snow.

Looking around he found that at long last he could see the top part of the door at the end of the Gordons' hut. He had walked a curved path in the snow, which had taken him right over the edge of the roof of the building and had fallen through the enormous drift formed by its protected side. It was now obvious why he had been unable to find the hut on his previous excursions: he had been walking a steady curve following the contour of the drift and not in a straight line at all.

With snow soaking his stockings and saturating his clothes, Jack slid and ploughed his way thankfully to the door. He tried to hammer on the top part but found that he was pounding on a fixed sheet of ice. Starting to clear the snow from the base of the door, he kicked and hammered but with no apparent response. He began to think that perhaps the Gordon Highlanders had left when the snow had started.

Finally, he heard a muffled 'och awa' wi' ya'. He shouted for someone to come to the door. With someone pulling from the inside, he tore thin strips of ice from the doorposts and, finally, with a frightful jerk, the door opened. Inside, a Highlander was standing in comical long woollen underpants and a long-sleeved vest. He had a blanket slung round his shoulders for protection from the cold and he eyed Jack through bleary, sleepy eyes. 'Whats a'do wi' mon? Awa' hame to your bed,' he said.

Jack pointed out that there were no beds and that one of the men was sick with hunger. Slowly the light dawned and suddenly realising the plight that they must have been in for the past two days and nights, the whole hut sprang into action. The canny Scots knew what to do in this kind of weather: they had simply hibernated. The snow had now almost stopped, and the morning sky had now taken on a lighter hue. Quickly preparing some cocoa and porridge on a coke stove in the centre of the room, some willing helpers helped Jack carry the welcome sustenance back to the CHL hut.

They clambered up the snow drift the size of which was now clearly visible. Its immensity in no way impressed the Scotsmen, but Jack stared in amazement at the huge dune-like snow drifts that stretched as far as the eye could see. He had never in his life seen anything quite like them in the south.

Though it had now stopped snowing, there was no way of getting down to the village, so they stayed put and awaited the next turn of events, after much argument and a host of promises to be carried out in the pub the next night.

Jack managed to borrow two bicycle chains from the Gordon Highlanders, and with these they got to work and repaired the transmitter turning gear so that when the line to Stanmore was eventually restored, they could get on with the war.

Jack became frustrated with the rumbling machine of Fighter Command. He complained that although he could see aircraft early on with his radar, by the time he had contacted Stanmore and they had contacted the aerodrome at Dyce, near Aberdeen, the German bombers could be well on the way back to base.

Ignoring the procedures that had been set up, he teamed up with some like-minded Czech pilots and devised a way of telling them his sightings before he told London. This involved setting up a secret phone line that went straight to the sector operations room and his friends at Dyce aerodrome. Using code words known to the Czech pilots, they would arrange for a section of Hurricanes to get airborne quickly, ostensibly on a cross-country exercise. Once airborne, the aircraft would fly in their direction. On reaching a reasonable altitude they could be seen clearly on Jack's radar well to the south. They agreed code words for the main rocky headlands and villages in the area. With the aid of a field telephone borrowed from the guard toom and connected to the transmitter modulation line, at the throw of a switch they could transmit to the fighter aircraft. The messages were short and sweet and must have been very puzzling for the men keeping the logbooks at Dyce. 'Apple, Angels Three' meant 'enemy aircraft in the vicinity of Fraserburgh were at 3,000 feet'. The method was successful intercepting several planes.

Inevitably, Jack's rule breaking was found out and he was arrested by military police and brought down in enforced silence to London and then to 60 Headquarters at Leighton Buzzard. Fearing the worst, he travelled down by train with his military escort, potentially facing a court martial or even execution!

On arriving at Leighton Buzzard, he went unpunished for his transgressions. On the contrary, he was rewarded for his ingenuity and asked to set up and run a radar station at Bolt Head in Devon; heaven to the 21-year-old. He had to arrange billets for about fifty staff and set up a Ground Control Interception (GCI) radar station on a clifftop, which had until recently been a field inhabited by cows. The station consisted of a series of trucks that contained the equipment. It was adjacent to an airfield and was exactly what Jack had been doing under the radar at Rosehearty and Dyce. The Cottage Hotel, which is still in Hope Cove, became the main billet, although there were more in Kingsbridge and Malborough.

While he was at Bolt Head, Jack commenced secret army training as he was keen to go on actual service. The Security personnel were often asked by their seniors to erase from their record log the reference to Jack being AWOL, because he had been given permission.

Eventually, in 1942 Jack was summoned to Whitehall to go on Operation Jubilee, the Dieppe Raid. It was a huge operation that was almost entirely manned by Canadian troops. The operation orders for Jack were that, because of his knowledge of Britain's secret weapon, the cavity magnetron and microwave radar, under no circumstances was he to be allowed to fall into enemy hands. This meant a bodyguard of ten Canadian sharpshooters were to follow his every move from the moment he stepped off the landing craft until his hopeful return to the beach to be picked up. The radar station was high on the French cliff overlooking the sea. It was going to be a massive task to get up there safely and back down again.

The raid was first planned for July 1942 as Operation Rutter. It was to be a large raid with full air power to protect the troops as they landed. Unfortunately, the convoy at anchor was discovered, bombed by the Luftwaffe and, consequently, the raid was cancelled. Operation Jubilee, the raid when it was resurrected, which took place on 19 August 1942, had less air cover and has gone down in history as a disaster. Tragically, many Canadians were killed or taken prisoner.

Before long, Jack's escort during the raid diminished. Eight were killed. He realised that it would be impossible to attack and enter the well-defended Freya radar station. He knew the Germans could see planes and ships on their radar and would inform the neighbouring station by telephone. Jack noticed there were many new telephone cables stapled to posts and trees. Entry impossible, he proceeded to cut every cable he could, forcing the enemy to use radio to inform the neighbouring stations, control centres and aerodromes what was visible to them. The moment the radio was used, the British on a ship in the channel recorded every signal. They could then analyse the radio messages and work out the efficiency of German radar and the defence communication system and develop counteractive measures to jam it. Some of this jamming was saved until D-Day, making sure the German radar did not see the massive forces about to invade France.

His job done, Jack scrambled back under fire to the beach and swam to the landing craft, all the time coming under fire. He managed to get on board with another member of the team. After the war, Jack became friendly with a German radar station operator, Willi Weber, who at that time could not understand why his phone had stopped working! Willi Weber later became a NATO General.

In 1943 Jack was posted to Egypt and spent the rest of the war travelling around the North African coast, assisting with the radar in the invasions of Sicily and Italy. Most of the time he worked in a large cave in Helman. He contracted malaria three times and became quite ill. He had a scar on his hand, which was the result of it being cut to release venom after a scorpion stung him while his hand rested on the floor. He was allowed to leave and went to Palestine. In his letters he was clearly struck by the contrast of cleanliness and Europeanisation after his prolonged stay close to the poverty-stricken villages of Egypt.

The war ended with a whimper for Jack. He eventually made it back to England and married Dally in 1946. Meanwhile, Jan Smuts's government in South Africa was seeking radar engineers to establish a network around South Africa. During the war, Prime Minister Smuts had visited Jack's radar station at Bolt Head. Jack was promised a fine job with prospects of promotion in a glorious climate and a new life. Likely feeling undervalued, here was an opportunity he could not afford to miss. He hoped for a better life in the sun and in 1948 the young couple decided to take up the position with the South African air force. Jack went out first and his plucky wife, now pregnant, travelled to Cape Town on the uncomfortable MV *Canarvon Castle*, which had been used as a troop ship and had still not been refurbished. Still, she had a lovely husband and a grand future waiting for them. What could go wrong? They were both looking forward to a good life in the pleasant climate, spacious accommodation and a well-paid job in the South African air force. They had a golden future ahead of them coming from post-war food-rationed Britain; it could only be good. Jack had fallen in love with South Africa when his ship had stopped off at Durban in the war on his way to Egypt.

Unfortunately, 26 May 1948 was a catastrophe for them, when the election was won by the Nationalist Party, led by Doctor Malan. In the air force it was now a must to speak Afrikaans. Newly arrived, Jack was told it was a prerequisite for his job. Undaunted and knowing some German and Yiddish, he mastered Afrikaans in three weeks; they gave him an exam which he passed. They were nonplussed, so gave him a supplementary exam asking him to name the bones of the hand in Afrikaans. They were determined he would not pass. He did not even know the names of the bones in English, so how could he know them in Afrikaans?

He was allowed to remain in the South African air force, but the golden job disappeared. His pregnant wife had arrived and so Jack was forced to take their offer of the lowest rank in the air force and therefore the lowest pay. There was no question of them coming back to England, even if they had wanted to. He had no money to buy himself out of a three-year contract or obtaining the return fare. Life was hard. They were transferred to Pretoria very short of money. As soon as Jack's three-year contract was up, he left the South African air force and opened a radio repair shop. This later became a hi-fi stereo factory.

The treatment of Jack became a cause célèbre; it was later brought up in the South African Parliament on 23 May 1950, and the South African Hansard records the response by the Defence Minister, who reiterated that all service personnel must speak Afrikaans. It also made the headlines in many of the newspapers.

Jack's shop was extremely popular, and he had time for everyone. All was going well for his family, but his life was about to get exciting once more, when he read a condensed book in the *Reader's Digest*. The book was *My Life in Court* by Louis Nizer, and one of the cases featured was the high-profile libel action taken by Quentin Reynolds against Westbrook Pegler, who had accused him of fabricating the newspaper articles about events in Britain.

Interestingly, while he was in Egypt in November 1943, Dally sent Jack a copy of the recently published *Combined Operations 1940-1942*, which had even passed the censor. She also sent him a copy of *Dress Rehearsal* by Quentin Reynolds, detailing his account of the Dieppe Raid. There are a few inaccuracies in the book, such as the number of men in the bodyguard, but it is obviously about Jack. In a letter to Dally, Jack talks about the book in an obscure way, but from the fact that she sent it, and him commenting on it, we can be sure they both knew exactly who 'Professor Wendell' was.

November 9, 1943

> That book has turned up and I've managed to read the more interesting parts. Incidentally Quentin Reynolds has written a book on that show in August last year. In it he refers to a mysterious person by the name of Professor Wendell and goes to great lengths to describe him as a wizard bloke in every way, but of course his real name in this particular case cannot be mentioned. As you know he was not killed, but I agree with him when he said that it's a good job the professor managed to get away. Even if he didn't have any trousers left. That's a story you don't know. [In reality, Jack lost most of his clothing in the swim to the landing craft.]

When Jack was touring around Canada with James Leasor in 1975 collecting material for *Green Beach,* James Leasor's book about Jack and the Dieppe Raid, it became clear that in Canada he would be appreciated as a hero. He was offered Canadian citizenship should he ever wish to emigrate from South Africa.

In 1977, Jack and Dell decided to leave South Africa and go to Canada, where they had a good life. It was here that Jack became involved in his third electronic career. This time it was in the field of museums and electronics, designing exhibits, especially the electric circuits, and training people to make hands-on machines for science museums all over the world, from Peking as it was then known, to Venezuela.

He also taught students at George Brown College, Toronto. The following appeared in *Mosaic,* the Student Service Department newssheet, in 1979:

> The Ontario Science Centre put Jack in touch with the college. He's been teaching color television for more than six months and enjoys his work. 'Color television uses almost every bit of today's technology, so you need a wide range of related experience. I love teaching. I'm a compulsive teacher,' he says.

Jack was well known in North America and was invited to be guest speaker at many reunions of The Rangers in the United States and various veteran associations in Canada. One particular invitation of merit was when he was asked to talk at NORAD, the secret aerospace defence facility deep underground in central Canada. They made him a name tag to wear bearing the words 'Professor Wendell.'

Don Preist and Jack met while working at Bawdsey. Their friendship endured for the remainder of their lives. In a video interview made in 1992, when Jack visited Don in California, Jack said that Don's work on Oboe, the British aerial blind bombing targeting system, had saved countless pilots' lives. Don then responded by saying that what Jack did in Dieppe saved countless lives on D-Day.

Jack's daring actions that day in August 1942 have now become a part of the Canadian history curriculum - a page is even dedicated to him in *Creating Canada,* the standard history textbook used in schools.

He died in Canada in 1997 and is buried in Pardes Shalom Cemetery, 10953 Dufferin Street, Maple, Ontario.

Jack felt very strongly that the contributions made by Combined Operations and radar to the success of the D-Day landings were not acknowledged, and in 1993 wrote the following:

> The magnitude and drama of the catalogue of events that took place between the apocalyptic eviction of the Allied forces from France in 1940 and their return in 1944 almost defy description. It could be that the writer's literary competency is not up to the task. His main qualification, which would possibly be unavailable to a more able author, is that he actually lived through a major part of this drama, much of which is not in the public record.
>
> Certain technicalities have to be described and explained if the full import of the miraculous events that took place on 6 June 1944, Operation Overlord/Neptune, are to be understood by the reader.
>
> Lastly, Earl Mountbatten of Burma played such an important part in moulding the organisation that brought about the victorious return to La Belle France, that some knowledge of his early personal history, particularly in the fields of naval engineering and communications, is vital.

In the narrative that follows, Jack describes in his own words how, in the meaning of the expression 'necessity is the mother of invention', the planning and creations of Combined Operations, coupled with the advances and brilliant use of radar, though often not acknowledged, made the Normandy landings as successful as they were.

CHAPTER 4
THE ROAD TO WAR

London: the year was 1932, and a typically cold dark damp miserable winter's day. It was one of the last sittings of the House of Commons and the Lord President of the Council, the Right Honourable Stanley Baldwin, closed the long and vehement discussion with a calamitous admission.

'There is no defence. The bomber will always get through.'

The House of Commons was in a state of shock.

Adolf Hitler had now been in power for just over a year and here was the Prime Minister of a victorious superpower just fifteen years after the conclusion of the 'War to End All Wars' admitting that the country would be powerless to defend itself against a new version of the Kaiser, Adolf Hitler. The estimated 5,000 bombers of Hermann Göring's air force threatened to rain fire, bacteria, gas and explosives on anyone who dared to pit their strength against the might of the Third Reich.

Gas masks were prepared, and in 1938 these were issued to every man, woman and child of a disbelieving and fearful Britain. The German dictator made it known that the Treaty of Versailles was defunct, and that Germany would rearm.

A new secret German air force, the Luftwaffe, had already prepared for a great onslaught on any country that dared to oppose the new German fighting forces, the Wehrmacht. All the public protest meetings came to naught, as there seemed to be no defence against Göring's bombers and Britain, despite all the sacrifices of the First World War, was once again at the mercy of a newborn Kaiser, the Reich Führer, Adolf Hitler.

In 1934, at the request of the Minister for State, the scientists of Britain quietly put their heads together and looked for a solution to the looming massive assault by Göring's estimated 5,000 bombers.

Headed by Dr Henry Tizard, a small select committee considered all suggestions and tried to come up with an answer to the threat of a mass bombing attack by the all-powerful Luftwaffe. It seemed that the ebullient 'abrasive' Dr Frederick Lindemann, a First World War scientist and great friend of Winston Churchill, was right. There simply was no electrical or any other method of detecting aircraft when they were many miles away.

In letters to *The Times* newspaper, the Professor advocated that his answer to coastal bombardment by German battleships during the First World War, still held good. When the German navy began to shell ports and coastal targets in that first 'war to end all wars', he had successfully advocated the mining of the approaches to Britain's ports and waterways.

As a senior scientist he demanded that his methods be applied immediately. All the talk about death rays to kill pilots and electrical methods of detecting aeroplanes, he insisted, was so much poppycock.

He wanted a barrage of huge balloons with mines suspended by steel piano wire to encircle British cities in order to dissuade any hostile bomber force from making mass air attacks on the towns of Britain.

His letters to the newspapers and correspondence became very much a crusade to waken the English people to the danger they were facing from the new despot in Germany.

By the end of 1934 it became obvious that this particular dictator was not going to go away. In fact, as Hitler secured diplomatic and military victories in defiance of the once victorious Allies, his hold on the people of Germany became stronger than ever.

CHAPTER 5
THE BEGINNINGS OF BRITISH RADAR

Dr Henry Tizard was quietly prompted by the Secretary of State to inaugurate a committee of enlightened scientists to establish an Air Defence System for Great Britain.

This select committee, which included Professor Blackett and Dr A.V. Hill, both with First World War military and naval experience, discussed what methods could be explored to protect Britain, which Churchill described as a 'big fat indolent cow waiting to be milked'. They called for suggestions and considered each and every one, but without any result.

Also searching for an answer on how to defend Churchill's 'big fat cow', the Director of Scientific Research at the UK's Air Ministry, Harry Egerton Wimperis, called on a semi technical Air Ministry official, Albert Percival Rowe, in an effort to resolve the impasse. They concentrated on the suggestion that perhaps a high-power radio beam could be generated that would boil the blood of a pilot, presumably spoiling his aim.

They had to find a radio expert who might be able to answer this very question, and the place to look was obviously the radio research division of the National Physical Laboratory. At their research laboratories in Slough, near Windsor Castle, was a man who, since the First World War, had been continuously engaged in scientific research, a canny Scotsman by the name of Robert Watson-Watt.

The question, in guarded terms, was put to him. Trying not to laugh, which is sometimes the most difficult thing to do when dealing with high dignitaries with all the power and no technical knowledge, Watson-Watt, with a straight face, offered to look into the problem.

He could have answered them there and then, but that would have been the wrong strategy. He broached the subject to his right-hand man Arnold Wilkins and in between chuckles they decided, as they knew exactly what the man was after, to keep a straight face and say that the 'hot blood' idea would not work, but there was a very good chance that they would be able to detect aircraft at a considerable range.

Albert Percival Rowe, the Air Ministry official, then asked Watson-Watt if he could set up a demonstration.

The Scot agreed and on one of the last days of a blustery and cold February in 1935, the three men met in a field at Weedon, near the BBC short-wave transmitter at Daventry. Wilkin's battery-operated, hand-assembled equipment had been installed in the back of a standard General Post Office telephone service panel van, and an arrangement had been made for a large aircraft, a Handley Page Heyford, to fly a course near the BBC Daventry broadcast radio transmitter and then over their heads.

Arnold Wilkins, the genius who was to be instrumental in setting up Britain's Chain Home radar equipment that was to bear the brunt of the action during the Battle of Britain, set up the apparatus the day before and slept the night in the freezing-cold General Post Office tender. Wilkins, working with Watson-Watt, used the BBC shortwave transmitter as a radar signal source and a van equipped with a cathode ray tube (CRT) receiver and antennae as the receiving station for the Daventry Experiment.

The next day Albert Rowe and Robert Watson-Watt arrived, and they all awaited the test flight.

Wilkins, who knew that the demonstration would work, was not particularly impressed, but Rowe and Watson-Watt were delighted when the trace on the CRT jumped up and down as the aircraft crossed the transmitted lobes of radiation: Britain had taken the first step up the ladder.

Immediately, £10,000 from a secret fund was granted to Watson-Watt and the game was on.

Using a site on Orford Ness, near Martlesham Heath aerodrome, on a site previously used by Henry Tizard during the First World War for his bombing experiments, the new tests were carried out.

They had immediate success with ranges on aircraft in excess of 30 miles and work went on at a hectic pace. A little way down the coast was the country home of Sir Cuthbert Quilter. An arrangement was made with the kindly old man, and, in no time, Bawdsey Manor became the home of Britain's first Chain Home radar station.

With Robert Watson-Watt as the superintendent and a senior staff of two, Arnold Wilkins and Edward 'Taffy' Bowen, they began to perform wonders, but of course all under the most secret security rating.

Meanwhile, Dr Lindeman was still persisting with his demands for aerial minefields around English cities. And, in another series of hard-hitting letters to *The Times*, he was accusing various politicians of dragging their heels with the dragon in Germany becoming ever more menacing.

Churchill, who knew of the successful test at Daventry, had the unpleasant task of bursting his friend's ego. Quietly, he had to ask his friend to stop insisting that electrical detection of aircraft was impossible, as it had now been achieved.

When the professor had recovered from the shock, he insisted that anything as sensitive as radio could easily be jammed and Britain would once again be left defenceless. He pointed out the fact that if you dropped 'dipoles', that is pieces of metal wire or foil cut to half the wavelength that the transmitter was using, they might each look like an aeroplane. The real bomber would not be distinguishable from the dipoles.

He was not very far off the mark, and the jamming problem was to become very real and frightening as Britain's Chain Home radar worked in the international short-wave broadcasting band and could easily be put out of action.

CHAPTER 6
CHAIN HOME RADAR

The initial development of radar was undertaken in 1935 at Orford Ness, but in February 1936 the research scientists moved to Bawdsey Manor House, where the stables and outbuildings were converted into workshops. There 240-foot wooden receiver towers and 360-foot steel transmitter towers were built, and Bawdsey became the first Chain Home (CH) radar station.

After a series of promising operational tests, the Committee for the Scientific Survey of Air Defence ordered the building of five of these stations to cover the area from the Wash to the Kent Forelands. More and very satisfactory tests were conducted with the mini chain, and as a result a whole chain of these stations, henceforth known as the Chain Home radars, was ordered to stretch from the Shetland Islands in the north to West Prawle on the southwest coast of England. It was known as the East Coast Chain and was to act as a shield if the Luftwaffe turned hostile.

Each CH station consisted of a massive transmitter built by Metropolitan Vickers of Manchester, a state-of-the-art receiver built by the A.C. Cossor company and a curtain array of aerials stretched between the four 360-foot steel masts, built by the Marconi Company, and four 240-foot wooden receiver masts to locate the direction of the returning echoes, as well as determine the heights of the aircraft under observation.

Throughout the ages, sailors have used a crude form of radar when close to rocky coastlines in misty weather. A sailor with a powerful pair of lungs would be stationed at the masthead. At regular intervals he would give a rather short, sharp yell. He would then listen for the echo. If it took a long time to return, he knew the ship was not in danger. If the echo returned quickly the dangerous rocks were nearby.

In radar, a short, sharp radio type of shout called a 'pulse' is used. The same theory applies: distant targets took a long time to echo back, and nearby targets would reply quickly.

Of course, radio travels at the speed of light and a suitably calibrated electronic television-type screen is needed to view the echoes (blips) and measure the time it takes the echo to return and hence the distance away of the reflecting object.

If you expect to receive an echo back from an aircraft 100 miles away, you would need a very powerful pulse transmitter. The Metropolitan Vickers Company, using BBC demountable tube technique, designed and built the transmitters and all the complex control apparatus needed to monitor the components once the transmitter was on the air.

The demountable transmitting tubes drew 120 amps in filament current. They did not last for ever and the design was such that when a filament 'blew', the transmitter was switched off and the transmitting tube taken to pieces. The thick filament could then be replaced and the process to switch the transmitter on again initiated.

Radio tubes had to be evacuated, for without a vacuum the filament would blow immediately. To suck out the air virtually down to the last molecule, a system of three vacuum pumps was automated. Initially a mercury vacuum pump was used and when most of the air had been evacuated, an oil diffusion pump automatically came on. When the vacuum reached a higher stage, another oil diffusion pump came on and sucked out the last vestige of air.

All this technology was monitored from a control desk. A red light would come on and the technician could then apply the tens of thousands of volts needed to operate the transmitter.

Being a very powerful transmitter, the tubes became extremely hot and would quickly have melted. To prevent this catastrophe with all the accompanying fireworks, a 1,000-gallon tank of distilled water was continuously pumped at high pressure through the valve jackets.

The whole transmitter was filled with 'fail safe' circuits to prevent an explosion. It took some time for the equipment to come on the air, and a skilled technician was always on duty at the monitor desk.

Once the transmitter was working, the radio power was fed out to the four 360-foot steel masts that supported the curtain array of aerials, which projected the radio 'shout' out for 200 miles and more.

The transmission was not a narrow beam but a kind of radio floodlight with a spread of 120 degrees. When the echo returned there was still the problem of determining from which direction it had come. This was the purpose of the 240-foot wooden masts, of which there were three. These were fitted out with sets of dipole aerials feeding into the operations room.

Here a member of the Women's Auxiliary Air Force (WAAF) rotated the knob on a direction-finding instrument invented at the turn of the century, the goniometer, and read out the compass bearing of the aircraft. Looking at the screen, she could also read off the range in miles and the height of the aircraft.

A plotter converted the range and bearing into a plot on a map, the coordinates of which were fed by an early online computer to Fighter Command Headquarters in London, where the aircraft was plotted on a table map for all the duty Command Controllers to see.

If a controller could lay claim to that particular aircraft, it was given a Friendly aircraft reference number. If not, it was given an 'X' number and fighter aircraft would be scrambled to intercept and identify. If thought to be Hostile, it was given a 'Y' number and when proven hostile, a 'Z' identification.

The main object of this discourse has not been to blind the reader with science, but rather to give some idea of the complexity and the enormous cost of a single CH station to the British taxpayer. With land appropriation, buildings, equipment and personnel, each station cost at least £250,000. In 1938 this was a very large sum for the taxpayer to bear and as the plans called for more than fifty CH radars, Professor Lindeman's criticism becomes quite real.

As if this were not enough, most CH stations had a smaller duplicate standby system installed underground in a bomb-and-gas-proof shelter. If the main CH station was destroyed, the station would still be able to function and pass plots to Fighter Command Headquarters at Stanmore and keep the ever-watchful controllers 'au fait' with the air situation in the area.

Radar jamming could make the whole terribly expensive investment absolutely useless. At negligible cost, a few enemy aircraft fitted out with simple 25-megahertz low power jammers could have rendered the whole Chain Home radar system worthless, leaving the men, women and children of Britain to the tender mercies of the Luftwaffe.

This was the quandary in the secret circles of radar at the time of the Munich Crisis of 30 September 1938.

That Easter of 1938, those CH stations that were operational came on the air on a twenty-four-hours a day basis. They were not to be switched off for many years. An hour was allowed each day for preventive maintenance, with an adjacent station always standing guard.

Quite obviously it was imperative that an answer be found to the problem of enemy jamming, and, after careful deliberations, it was considered that there could be four main types of radar jamming.

Three were answered with counter-countermeasure circuits built into the receivers by the Cossor company. For the fourth and worst type, 'white noise amplitude modulated continuous wave', there appeared to be no counter.

For almost two years this frightening prospect hung like the sword of Damocles over Robert Watson-Watt's head. For a ridiculously low investment, the Luftwaffe could put Britain's air defence system completely out of action. It was only when, on 3 September 1939, war had been declared, that the Mazda Company finally came up with the answer. It took the form of a special long persistence afterglow memory-type television screen.

Manufactured by the Mazda Company, it had multiple phosphors in layers. The rear phosphor layers filtered out the unsynchronised 'hash' of the jamming and allowed only the steady repetitive synchronised aircraft blip to get through to the front screen.

Some hash would also reach the front screen, but as the colour of the hash and the blip were different, optical filters which could be placed over the radar screen thankfully made the equipment serviceable.

It had taken some years to come up with the answer and an Anti-Jamming type screen was kept under top secret cover on every CH station, only to be used when the stark necessity arose.

Fortunately, all this difficult research and development took place when we were still at peace. General Wolfgang Martini, the Director General of Luftwaffe Signals, was to be permitted no such luxury.

We had won by a short head.

The ultra secret code name for this vicious type of jamming was 'Mandrel'. It was to serve us well.

In spring 1938, at the time of the Munich Crisis, those CH radar stations that were serviceable came on air. They were only finally switched off in 1946, some eight years later, at the end of the Second World War.

Now with a secret protective fence stretching out more than 100 miles from the Shetland Islands in the north to West Prawle off South Devon, the Air Ministry felt that Britain's air defence was secure, but it improved with each new CH station that was commissioned.

Though the radar worked magnificently at long range and at the normal operating heights of possible bomber aircraft, it was quite blind in respect of low flying aircraft.

Aircraft approaching at less than 3,000 feet would not be detected until they were within 30 miles of the station. This serious shortfall was remedied by preparing a smaller radar working on a much shorter wavelength. As this station would be able to detect low flying aircraft and would be part of the Chain system, it was codenamed CHL or Chain Home Low. With a pair of these tiny stations, one on each side of the CH station, it was possible to detect all aircraft down to about 500 feet above the sea.

At the time of Munich, the men of Bawdsey had produced a number of devices all intended to improve Britain's defensive capabilities if and when war was declared. A vital auxiliary device was the IFF unit, or Identification Friend or Foe. It was a tiny 18-inch square box (the British 12-volt receiver transponder unit type R3108 IFF Mk II weighed 11 kg and measured 210x280x300 mm) to be carried in all friendly aircraft, and it would constantly signal back to all the CH radars that this was a friendly aircraft that was under observation.

The first unit was prepared by Dr Donald Preist, one of the original Bawdsey team and a close friend of the author, about whom we will read later in the story.

The device is now known internationally as a transponder and is a vital part of all our operations in space as well as in satellite communications.

Its mode of operation was quite simple. It received the powerful pulse coming from the CH transmitter and immediately sent a pulse back to the CH receiver screen. The CH operator, as usual, could see the echo or blip from the aircraft in the normal manner, but now the recognisable IFF pulse was superimposed on it.

Initially for his experiments, Don Preist used a full-sized receiver and there was a delay between receiving the CH pulse and its transmission back to the CH. For this reason, on the screen the aircraft blip was a mile or two away from the IFF pulse, making it seem that there were two aircraft under observation.

Partly by using a simple receiver this effect was overcome, but not before Don Preist realised that this could be the basis of a very good radar countermeasure. Later in the war it was developed, code-named Moonshine, and first used two days before the Dieppe Raid and then very successfully on D-Day.

After Don Preist had proved that his IFF system would work, Watson-Watt placed burly Bob Carter, another of the original Bawdsey team, in charge of miniaturising the apparatus and developing a standard small IFF unit that could be carried in a fighter aircraft.

Preist was then given another and very difficult assignment. Another Bawdsey invention produced by Edward 'Taffy' Bowen was a radar small and light enough to be carried in an aircraft. It was code named ASV or Anti Surface Vessel radar.

Some years before the war, this radar had alarmed the Lords of the Admiralty when it located major units of the Home Fleet under what would normally be impossible conditions. This occurred when the navy was forced to call off an exercise in the English Channel due to heavy fog. Without radio on their aircraft and not knowing of the cancellation of the exercise, Edward Bowen located the aircraft carrier *Courageous* and the other major units of the fleet, and using a public telephone reported their position to Coastal Command. This caused much consternation at the Admiralty, but no particular action was taken by the navy as a result of this obviously revolutionary development.

This wonderful and simple radar could spot capital ships at ranges of 25-30 miles and was to play a large part in containing the U-boat threat.

The ASV also plays a large part in this saga. Very much the same equipment, suitably modified, it became our standard Aircraft Interception radar or AI, which was instrumental in shooting down many German bombers at night.

Apart from the GL (Gun Laying) radar, an anti-aircraft gun laying system, these then were the main radar devices with which Britain entered the war.

At 1 o'clock on 3 September 1939, and to the sound of the first wartime air raid siren, war was declared. The transition from peace to war, which took place with the Prime Minister's announcement, was painless.

The fact that we were now at war was only brought home to the people of England because the air raid sirens immediately went into action.

A French aircraft, innocently approaching the Thames estuary, had been plotted by the CH station at Canewdon and, not being fitted with IFF, the rules of the game were now that it would have to be intercepted and verified as friendly or hostile.

CHAPTER 7
DUNKIRK

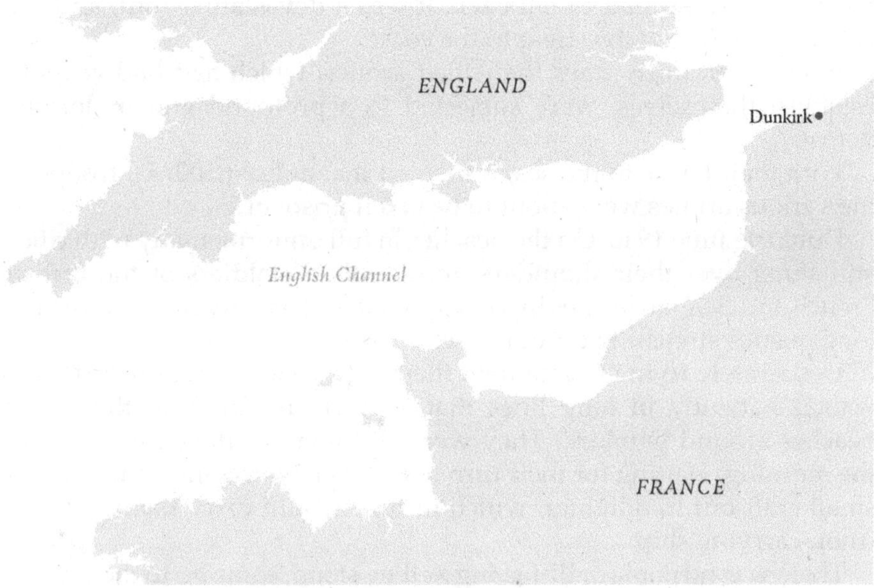

Dunkirk, May - June 1940

The atmosphere in England was almost one of disbelief, as life went on as usual. Certainly, more and more soldiers were called up and sent to join their friends already established in France and manning the borders between Germany and France.

France had its Maginot Line of defence, and Germany had its opposing Siegfried Line, and both were the butt of many music hall songs of the time.

In France and Belgium, where the troops were stationed, the girls were pretty and life for these ex-civilian soldiers was more than bearable. This was the period of the 'phoney war' and the build-up of troops guarding the borders with Germany escalated until by

May of 1940, some millions of French, Belgian and British troops had established themselves in defensive positions.

In April, in a surprise action, the Wehrmacht occupied Norway.

In May the Germans invaded Holland and Belgium, and the battle for France commenced.

Using their highly developed 'Blitzkrieg' technique, German armoured columns outflanked the Maginot Line and approached France in the north through Belgium.

In the south armoured columns penetrated the Ardennes Forest and the Allied armies had to fall back. It was a devastating rout, and the armies were forced to retreat to the coast.

In less than thirty days the Allied armies, which had had years to establish themselves, were subjected to a profound and inglorious defeat.

With their backs to the sea, it seemed that at least 500,000 troops of the various armies were about to be taken prisoner.

Dunkirk: June 1940. On the beaches in full uniform, many with rifles still slung over their shoulders, thousands of soldiers of the British, French and Belgian armies in a disciplined and orderly manner formed long queues stretching far out to sea.

Desperately trying to maintain their balance in the rough surf, they waited patiently in long lines that snaked far out from the sandy beaches around Dunkirk. They were still there in the early hours of the morning, waiting for their turn to be taken by rowing boat or other small craft out to a lighter, which in turn would carry them out to a troop-carrying ship.

There was no panic or ill-feeling as they stood, some up to their necks in sea water. Old and young, they did not want to lose their position and possibly the chance of escape from the overcrowded Dunkirk beaches. With the Dunkirk mole heavily damaged and piled high with wrecked vessels of all kinds, the beaches were the last chance.

Getting aboard ship was not necessarily the end of the story, however, as they had all seen many fully loaded troop-carrying ships bombed and sunk just offshore, drowning thousands. Their bodies were strewn along the surf line on the beaches.

Apart from the air bombing, there were also the mines being laid by the German aircraft, which took a steady savage toll of the ships. The backdrop to all this carnage was the incessant land-based artillery, pounding, killing and maiming the luckless soldiers. Many heavily loaded ships that almost made it back to England were torpedoed by the watchful, waiting U-boats that were patrolling the Channel.

At Admiral Sir Bertram Home Ramsay's request, enormous fleets of small craft had left 1,001 English rivers, ports and harbours in an effort to assist in the rescue of the defeated battered remnants of the Allied armies.

With almost half a million men concentrated in and around Dunkirk and its beaches, which at this time should have been crowded with happy holiday makers, the all-victorious German Luftwaffe, army and navy had a heyday.

They could shell, machine gun, torpedo, mine and bomb their victims at their leisure. The beaches themselves were pockmarked by 1,000 bomb craters and some areas were red with the blood of the more unfortunate soldiers. The mangled remains of soldiers of the defenceless British and Allied armies littered miles of the coastline of northern France, together with the wreckage of large numbers of ships.

For almost two years the British army, navy and air force had poured men and millions of tons of equipment into France to face the German army which had sat smug and secure in the 'embattlements' of the Siegfried Line, a line of strong points that separated France and Germany.

Though war had been declared more than six months previously, there had been little or no action for the British and French soldiers facing the Germans.

In their own underground bomb-proof observation defence centres, the High Command of the Wehrmacht planned its surprise assault.

At sea there had been a considerable amount of action. Ships had been sunk and the approaches to the Port of London had been heavily mined by German aircraft, which had already claimed many casualties, as was evidenced by the forlorn wrecks of ships resting on the shallow mud banks of the Thames estuary.

Hitler had not wanted to go to war with Britain, and in radio broadcasts, which had been a mixture of threats and cajoling, had tried to get Neville Chamberlain, the British Prime Minister, to come to terms and call off his declaration of war with Germany.

The German dictator felt that he could handle Chamberlain, who had always fallen in line with his demands in his vain efforts to preserve peace. Their talks had continued on and off for about two years, starting before Britain's sacrifice of Czechoslovakia in 1938, with which Britain and France had signed a pact of mutual defence.

Chamberlain's policy of appeasement to Hitler has been a depressingly disgraceful episode in Britain's history. With a supposed guarantee of assistance from both Britain and France, the powerful Czech army was ready in 1938 to settle Hitler's hash once and for all, but France, not wishing to antagonise Hitler, together with Britain had capitulated to Hitler's demands and Czechoslovakia was thrown to the Nazi wolves.

The Czechs, with no help from their supposed allies, had had to acquiesce in what Chamberlain had been assured would be Hitler's last piece of international brigandry.

With the powerful Czech Skoda arms works now under his control, together with the highly mobile armoured corps developed by the Czechs, Hitler for the first time had a fighting force no longer dependent on bluff and more than capable of putting the theory of German Blitzkrieg into practice.

Already in April 1940, the German army had invaded and occupied Norway and Denmark, securing what Hitler regarded as his northern flank. Hitler still hoped that he could make Britain come to heel. But a new Prime Minister, Winston Churchill, had taken over the reins, and he refused to negotiate with the man whom he referred to as Corporal Hitler.

Suddenly, in May of 1940, the German army went on the attack. Initially they devastated Rotterdam, and invaded Holland and Belgium in a successful attempt to break the Allied northern defensive line. They knew that the British would be honour bound to come to the aid of the Low Countries.

This feint in the northeast worked admirably and, when the British were totally embroiled in the defence of tiny Belgium, the German tank columns suddenly achieved what General Maurice Gustave Gamelin, the Commander in Chief of the French army, had stated was utterly impossible: they had crashed their way through the forests of the Ardenne in southeast Belgium, and in a carefully planned amphibious attack had punched their way across the River Meuse and into France.

Against orders of the High Command, the German army, under the command of General Heinz Wilhelm Guderian, making their way in a wide sweep, scored an astonishing victory, quickly reaching the coast of northern France. Using their Blitzkreig (Lightning War) technique, their armoured columns advanced at high speed, only calling in the dive bombers when they met resistance.

Using their top-secret Enigma radio communications system, which permitted excellent coordination of troops and aircraft, the advancing Panzers smashed all resistance. Using modern radio communications, the Germans were fighting a state-of-the-art war completely outclassing the Allied armies. Still fighting a 1918-style war, General Gamelin, pampering himself, lived and dined in luxury at his headquarters in beautiful Vincennes. With no radio communication, all dispatches to his troops in the different theatres were conveyed by motorcycle. As the Allied forces crumbled under the German attack, the French army took to the ditches with the terror-stricken civilians in order to escape the howling Junker dive bombers.

They had every reason to shout 'Treason', as for many years the officer class of the French forces had been under the increasing influence of a right-wing organisation known as the 'Croix de Feu'. To these men, Hitler was the leader who was going to bring about a new order in Europe, if not the world. Secretly, they co-operated with the Nazis and did not want to resist Hitler's plans, in which France had been promised a major role. This 'fifth column' perhaps explains the complete inertia and paralysis in certain areas of the plan for the defence of France.

Before the attack, the Deuxieme Bureau de l'État-major Général (Second Bureau of the General Staff) had, in great panic, reported a gigantic build-up of Wehrmacht tanks and troops on the roads to the Ardennes.

The French air force did not carry out even one reconnaissance flight to verify this vital intelligence information.

As General Guderian's armoured columns pushed north at high speed through the Low Countries, the British had no alternative but to fall back to the coast of France, hoping for an evacuation operation that might get them back to England.

Fighting bravely in desperate rear-guard actions, the remnants of the British Expeditionary Force were gradually compressed into a small defensive perimeter around Dunkirk, keeping the rampant victorious Germans at bay as they delayed the final and inevitable onslaught.

Dunkirk had quite a sizeable mole that could accept fairly large sea-going vessels, and it was operating round the clock moving huge numbers of troops.

Apart from aircraft that incessantly bombed the ships and their prospective passengers, the Germans brought up medium-range artillery, which shelled the beaches, sinking ships and making life miserable for the sailors and the would-be escapees.

The situation was disastrous and with no chance of survival, the troops set to work destroying millions of pounds worth of military equipment in massive beach bonfires.

Thousands of military vehicles were also destroyed to prevent them falling into enemy hands, as the troops made their way down to the beaches hoping for a miracle.

This was in fact what was needed, if the men were somehow going to be lifted and brought back across the English Channel.

As we know, Admiral Ramsay wrought a 'miracle' by enlisting the aid of the skippers of hundreds of small river craft in southern England, many of whom had never really been out to sea. The armada of small craft made their way across the English Channel and in the face of heavy artillery fire, air bombing and machine gun attack, they ferried the waiting soldiers out to the larger ships.

The losses were severe and included fourteen destroyers and countless other ships of the merchant navy, but a large proportion of the army was saved to live to fight another day.

In these few pages, it is quite impossible to visualise or describe the magnitude of this miraculous transportation of more than a third of a million souls from France to England in a matter of days.

Perhaps the recital of some figures taken from the official list of ships involved in the Dunkirk operation might give some idea of what had taken place.

From the rivers and seasides of Britain there were 520 miscellaneous small craft, 45 destroyers of the Royal Navy, 20 warships of the French navy, 50 various ships of the Royal Navy, 39 minesweepers, 40 minesweeping vessels of various types, 20 anti-submarine trawlers, 12 motor torpedo boats, 20 beach landing drifters, 12 Portsmouth yachts, 40 miscellaneous Royal Naval craft, 45 Dutch coastal freighters, 45 troop carriers and hospital ships, 20 merchant navy vessels and 35 tugs.

Amongst the many ships lost at Dunkirk was the River Thames pleasure boat the *Crested Eagle*. It was on this small tourist pleasure craft that the writer, as a young lad, was first introduced to the sea. When about 5 years old, my father had taken me on the 'voyage' from the Tower of London to Ramsgate on the North Foreland of the Kent coast. Strangely, in 1940, I was at CHL Radar Foreness, when I heard the sad news of the demise of this magnificent and colourful vessel that had given us so much pleasure.

I understand that the two sister ships the *Royal Eagle* and the *Golden Eagle* were both damaged but survived this stupendous but thankfully successful evacuation operation.

This spontaneous unthinking support by the men of the small boats augured well for the future. It was a victory of ardour, which passed quite unnoticed by the all-victorious Germans.

In the true Nelsonian spirit, coercion was not needed. Every man, young and not so young, did his duty, and sadly many paid the supreme sacrifice.

The victory of the German army was all the more amazing as it had taken place in one short month and in face of 100 divisions of Allied troops ensconced in carefully prepared defensive positions. These had taken years to build at the cost of millions of pounds and as many tons of concrete. In a few short weeks this had all been negated by a small, armoured force of the German Wehrmacht.

Having been completely defeated and summarily ejected from continental France, the prospect of our ever being able to dislodge the Germans once they had established themselves there seemed utterly remote.

Hitler and his cronies demonstrated their elation on the German Propaganda Ministry's newsreels at the time, and the euphoria spread throughout the victorious German forces, many of whom were sent home to Germany for demobilisation. The war was over. Germany had won. At least that is how it looked to the people of France.

In France, the German soldiers, whose behaviour was correct and restrained, became quite popular with the girls. In the brilliant summer of that victory, the shops of Paris reopened and enjoyed a boom time as the smiling Germans sent silk stockings, scarves and perfume home to their wives.

Summer 1940 was wonderful and perhaps this had a positive effect on the morale of the people of Britain. Though shocked at the destruction of the Allied armies in a matter of weeks, there was still no fear. This despite the fact that the all-powerful Wehrmacht seemed invincible, and that even if our soldiers could be brought back to England, we would immediately be faced by an invasion of the mighty German army, supported by its gigantic murderous air force, using airborne troops in daring operations, which had been crowned with success particularly in the Low Countries. The pattern of things to come was all too obvious to the people of England.

The last line of defence seemed now to be the heavily battered and sadly depleted Royal Air Force. The small fighter force of perhaps 600 first line planes would soon be disposed of by the 2,400 first line fighters and bombers of the Luftwaffe and the invasion of Britain would then most surely follow. In typically efficient manner, the Nazi hierarchy had already drawn up plans for the administration of a defeated Britain with gauleiters already appointed to take charge of civilian government in southern England.

In Britain, many well-known personalities had for years consorted with Adolf Hitler and had made no secret of the fact that they admired the man and his policies. These men were expected to now come out into the open and take up the appointments offered by their new master.

CHAPTER 8
THE BRITISH RADAR SCENE

As we know, there was, however, another line of defence about which the people of Britain knew practically nothing. It was the device and organisation that would magically transform Britain's 600 fighter aircraft into 6,000.

It was radar in the form of the CH, the CHL and IFF in our fighter aircraft that was to prove our salvation in the trying months ahead by multiplying the effectiveness of our tiny fighter force.

The CH station could determine the position and height of an aircraft providing that it was flying between 3,000 and 25,000 feet. However, at lower altitudes its performance fell off. In fact, below 3,000 feet the station was to all intents and purposes blind. As previously discussed, to overcome this very serious defect another radar system was devised that could see aircraft down to 500 feet from the sea. These were called Chain Home Low (CHL) stations.

Working together the two radars gave Britain adequate coverage against surprise attack. And IFF, now universally known as the transponder, had been prepared that indicated to the ground controllers if the aircraft under surveillance was in fact friendly or not.

A further radar device for anti-aircraft gun control had been devised and was known as GL or Gun Laying. It was not a great success. There were also two tiny airborne radars: the ASV (Air to Surface Vessel), which could detect large ships at a range of 20 miles or more and the AI (Aircraft Interception), a night fighter radar that allowed the pilot to locate an enemy bomber under night conditions.

The detail of the whole integrated radar defence organisation was shrouded in great secrecy, and it was this that was to prove to be the decisive weapon against the Luftwaffe.

Another deadly secret was the fact that in January 1940, two Birmingham University lecturers had produced the ultimate in microwave transmitters, the Cavity Magnetron.

After years of fruitless search, the scientists of the United States, Germany and, for that matter, the rest of the world, had come to the conclusion that for cogent technical reasons, the shortest wavelength on which radio equipment could transmit at high power was 0.5 m. Furthermore, for other technical reasons, the shortest wavelength on which radio receivers could effectively operate was also 0.5 m.

As a result, all German secret development of radar was based on the false assumption that it was utterly impossible to develop radar equipment that would work on this very short wavelength of less than 0.5 m.

With typical British obstinacy, two young lecturers, John Randall and Harry Boot, working with their students at Birmingham University, had developed and manufactured a super high-power transmitter tube that worked on the incredibly short wavelength of 10 cm. They did not know at that time why it developed such high power, but that was of little consequence, as it most certainly did. Strange to relate, the basic magnetron oscillator was a German invention. The Englishmen had merely gone one small step forward and produced the Cavity Magnetron, a device that has transformed our lives to this very day.

Their phenomenal discovery had to be kept secret, as it was clear that when developed and used in radar and radio communications apparatus, it would undoubtedly revolutionise every aspect of the science of war. Its powerful microwave transmissions heralded radar equipment of incredible precision, making our CH and CHL systems seem archaic.

Of course, much water would first have to flow under the bridge before these microwave radars would see the light of day.

CHAPTER 9
THE GERMAN RADAR SCENE

Following the capitulation of France, the German occupation took place immediately. Under the strictest security a strange secret item of military equipment arrived at the coastal village of Pourville, near Dieppe.

It was a German early warning radar, codenamed 'Freya', which at this time was mounted on caterpillar tracks. In appearance and performance, it was strangely similar to our CHL. Placed on the clifftop site it could plot aircraft up to a range of 50 miles. With its 20-degree beam sweeping like a lighthouse, as did the CHL, it was able to plot large ships up to a distance of 25 miles. Like the CHL it had no great precision, but it was more than adequate for daylight operations when a fighter aircraft could easily see his target from a few miles away. The range indication was very precise, but the method of reading the bearing of an aircraft was subject to an inherent error, and this could be quite a few miles at its maximum range.

Freya radars were towed into position and installed at other strategic points, including Wissant near Calais and Cap de la Hague, Cherbourg.

The second type of German radar, codenamed 'Seetakt', was the type found aboard the battleship *Admiral Graf Spee,* which had been scuttled on 17 December 1939 by its captain in Montevideo harbour, Uruguay.

At this time the Royal Navy had no fully developed shipboard radars, and this was indeed an unpleasant surprise to our scientists. This was our first indication of the fact that the Germans had also invented radar.

The Seetakt was a precision gunnery radar and a superbly engineered instrument. Mounted ashore on a fairly high cliff, ranges of almost 20 miles could be expected on large surface vessels.

Apart from the Freya and the Seetakt there was a third excellent gun laying radar, codenamed 'Wurzburg'.

At that time, we had no gun control radar to compare to the Wurzburg, which technologically was years ahead of its time. Its range performance was only about 12 miles, but when used for directing anti-aircraft guns or searchlights, it was a superb and very dangerous weapon, as the men of Bomber Command were to discover.

On their arrival at Calais, the Germans immediately set up long-range cameras to photograph the cliffs of Dover, which were clearly visible across the straits.

We had received a shock after finding the Seetakt radar aboard the *Admiral Graf Spee*. Though it had been a severe jolt to our ego when we discovered that the Germans had also invented radar, it had no immediate effect other than that it was a very important piece of intelligence information. It enhanced our knowledge without doing any particular harm to our defence capabilities at that point in time.

On the other hand, the German discovery that Britain did indeed have radar came as a profound shock to them and at a most inconvenient time. With only weeks to go before the launching of 'Adlertag' (Eagle Day), the air assault on England, the discovery was a serious tactical blow to the Luftwaffe. It suggested that the air attack might be more costly than the attacks on the defenceless cities of Guernica, Warsaw or Rotterdam. It could make life difficult for the German pilots, as despite the series of prewar spying flights around England by their Zeppelins, they had not discovered that we had a superb integrated defence system in place.

As a result – and this was to prove an unmitigated disaster – General Wolfgang Martini, the Director General of Luftwaffe Signals, had not been able to order the design and manufacture of the appropriate radar jammers.

The few weeks left before 'Adlertag' on 13 August 1940 did not permit the design, manufacture and installation of jamming equipment into German aircraft.

This one vital omission was to cost the Luftwaffe an easy victory in the forthcoming Battle of Britain.

CHAPTER 10
GROUND-CONTROLLED INTERCEPTION

Even before the drama of Dunkirk had been played out, the Chief of Fighter Command, Air Chief Marshal Hugh Dowding, prophesied that if there was a big air battle over Britain, we would triumph. The Germans would then try to attack by night. For this reason, he requested from the radar community that they give thought to the production of a radar that would be able to operate inland and control night fighters over the English countryside.

Very quickly, a unit christened GCI or Ground Control of Interception, which was a cross between the CH and CHL, was devised and training commenced immediately at Westhampnett, near Tangmere in Sussex. This new night-time radar could be installed at inland sites and could read the heights on all aircraft and perform all the necessary air navigation for the night fighter pilot. With this responsibility removed from his shoulders, the pilot could then concentrate on the onerous problems of flying and fighting in the dark.

Using a Bristol Blenheim aircraft fitted out with the Aircraft Interceptor (AI) radar, developed from scratch by a Welshman, Ed Bowen, they managed to shoot the first hostile aircraft down at night.

All this happened in August 1940 and for many reasons, the initial success could not be easily repeated, but development proceeded, and six GCI mobile units were ordered.

By the end of 1940 the Blenheim was replaced by a more suitable aircraft, the all-metal Beaufighter. With a higher turn of speed, more engine power, better weaponry and a more predictable radar transmission characteristic, it became our standard and most successful night fighter.

However, the AI radar as installed on the Beaufighter had one serious defect. The last half mile of the approach to the bomber was blotted out by the large pulse transmitted by the AI. The pilot could not get near enough to the target bomber to carry out the visual identification needed before opening fire.

This last problem was finally solved by Alan Blumlein, one of the most prolific inventors of the twentieth century, who worked as a civilian engineer at the EMI prewar television factory. Blumlein's modulator was installed in the Al radars and immediately the night successes mounted.

In January 1941 three bombers were brought down at night. In February twelve were shot down and the figures mounted as the six new GCI radar stations came into service.

The stations developed a 'handover' procedure, where the target was passed like a rugby ball from station to station and the kills multiplied. In fact, in the last month of heavy bombing, which terminated in the first week of May, almost 100 bombers were shot down at night.

This was too much even for the mighty Luftwaffe and their heavy attacks to all intents and purposes ceased. We did not know why at the time, but as we now know, Hitler had decided to look east.

It was exhilarating to be engaged in a new science which was top secret. This allowed those involved in radar work to constantly come up with new applications of radar theory.

One such, which was to prove of great value to this very day, was based on the use of the IFF unit, not in the aircraft but on the ground. Mounted complete with a battery and a 10-foot-high aerial at the end of the runway at a GCI station at Bolt Head, Devon, it enabled aircraft based there to locate themselves in the night skies for the first time.

The AI set in the night fighter could clearly see the IFF on its own radar tube at its correct range in miles to the aircraft. The pilot could then turn his aircraft in the direction of the IFF using it like a beacon and find his way home. Also, if it was placed at a known point at the end of the runway, it became an excellent landing aid for the night fighters. Another bonus was the fact that two different IFFs located at two separate aerodromes would give the two ranges in miles. By describing two arcs on a map, the pilot knew that he was at that spot where the two arcs intersected. This was indeed a great breakthrough in airborne radar navigation. For the first time the pilot knew exactly where he was, despite the black void of night. Individual ground-based IFF units were now referred to as beacons and were used by the night fighters as orbit beacons. They could, with no help from the ground, circle a beacon as if on a taxi rank until required for the interception of a German aircraft.

The IFF, now called the 'transponder', is used throughout the world in many ways by airlines, satellite communications and in space projects, where it has become a workhorse.

Throughout the Second World War, new radar systems were tailored to fit problems that had hitherto been regarded as insoluble.

As we will see, this was especially true of the new fighting service, Combined Operations Command, the organisation set up to prepare the ships, men and equipment that would one day, we hoped, allow us to return to the continent of Europe.

Sadly in 1941, thoughts of our ever returning to France were not really a priority.

We had first to survive that which was constantly on Churchill's mind, the dreaded German invasion.

CHAPTER 11
HOLDING THE LINE - THE BIRTH OF COMBINED OPERATIONS COMMAND

During the First World War, when Britain was being bombed and battered by the Zeppelins and Gotha bombers of the German air force, Field Marshal Jan Christian Smuts had been appointed chief of Britain's Air Defence. This wonderful old soldier, who had led the Boer armies against the British during the 1899-1902 Anglo Boer War, was now a close friend and advisor of Winston Churchill.

Churchill had in fact been captured by the Boers and had managed to escape from his prison in Pretoria and rejoined the British forces.

When Lord Louis Mountbatten was a young lad, Churchill regaled him with his adventures in Africa during the Boer War, building into the young man the fact that leadership, guts and initiative alone would win victory in wartime. Even more certain, sitting on your rump would not.

At the turn of the nineteenth century General Smuts had led a tiny nondescript guerilla army against the British forces during the Boer War. Time and time again he turned defeat into victory.

To keep the fighting spirit of his men alive he sent tiny groups of tough fighting men into the British camps in the dead of night to terrorise the British invaders by cutting the throats of soldiers on night guard duties. His men would blow up ammunition dumps and steal stores, arms and equipment from the invading armies, causing alarm and despondency amongst the British troops out of all proportion to the numbers of their adversaries.

These dedicated brave men who were expert shots could live for weeks on a very nutritious dried meat, 'biltong', which they carried in their saddlebags. Tough, unbeatable guerrillas, they carried the war to the British, whom they claimed were usurping their land. They were known as the Boer Kommandos.

In June 1940, immediately after the Dunkirk debacle, Churchill once again called on his old friend for advice, guidance and uplift in his time of need. General Smuts told Churchill to immediately form Kommando units to murder, harass and terrorise the German troops that were in the process of occupying France.

Many of the men only recently ejected from France, having lost brothers, friends and relations in the process, were burning for revenge at any cost. These highly motivated troops were to rebuild the morale of the British army after its humiliating defeat and keep alive the spirit of resistance to the Nazi invaders.

With the French on their knees, the Germans were now joyfully reaping the benefit of their victory and taking up defensive positions along the coast of northern France.

Smuts said there was no time to be lost, and Churchill issued the orders immediately.

On 3 June 1940, when British troops were still filtering out of western France, Churchill issued a minute to the Chiefs of Staff asking for the small formation of tough troops to be formed to carry out spiteful raids on the coast of German-occupied France: 'The completely defensive habit of mind, which has ruined the French must not be allowed to ruin our initiative . . . Enterprises must be prepared with specially trained troops of the hunter class, who can develop a reign of terror in the first place on a "butcher and bolt" policy.'

Within weeks of the Dunkirk defeat, small groups of British soldiers were landing in France at the dead of night, attacking German outposts and building up a murderous reputation amongst the troops now occupying the conquered territory.

The Wehrmacht learned to hate and fear these men, whom Churchill, in deference to General Smuts, had christened the Commandos. These magnificent, superb soldiers were to gradually grow in numbers and spearhead all the major Allied assaults on enemy-occupied territory.

The Commando Brigades, the Royal Marines and the Marine Commando, together with the men of the Airborne Divisions, many specifically trained for amphibious operations, were to be the cutting edge, the fighting men of a new arm of the British services, Combined Operations Command.

Churchill asked his friend Admiral Sir Roger Keyes, a First World War hero, who had successfully raided enemy-occupied territory in the conflict, to set up the new command.

An experienced and very senior officer, Keyes immediately located premises at 1A Richmond Terrace in Whitehall, within shouting distance of the Admiralty, with whom he would have to 'cooperate'. He took over a portion of the building which had recently been the headquarters of London's Metropolitan Police.

His Cabinet directive was to build up an amphibious fighting force to carry out raids on the enemy-occupied coast and ultimately to invade the continent of Europe. In July 1940, this was little less than a pipe dream.

There was no shortage of volunteers from the survivors of decimated British formations, and mini attacks started very quickly. The spirit of some of the 'tea-time' soldiers had also undergone a drastic change. Certainly, the vast majority were still determined not to 'put themselves in harm's way' if they could possibly avoid it.

A sizeable number of highly motivated volunteer troops were drawn from their regiments and given the appropriate murdering skills in tough training establishments. Though they would wear their own regimental badges, they were attached to small units and were billeted in boarding houses in coastal areas, where they could train in beach landing exercises.

When Major General Bernard Montgomery was offered the command of these tough characters, he disdainfully declined. He did not believe in 'prima donnas' and disapproved of the irregular guerilla kind of operations they intended to conduct.

The Commandos were to upset Hitler, who in desperation issued a decree ordering their summary execution on capture.

Volunteers from other elite formations such as the Guards and the Highland Divisions moved into the Commando and, despite the Montgomery attitude, the Commandos were to expand into a powerful fighting arm with an impressive list of glorious victories to its credit.

But for the ordinary soldier, there was none of the 'do or die' mentality; they wanted to go home as soon as possible. When the first Commando units were formed, volunteers were accepted from all regiments. The preparation was more than tough, and many dropped by the way, but the training had to be very rigorous, as only dedicated fighting men were wanted.

When in 1941 Sir Roger Keyes chose his men, apart from the naval and marine specialists, they all came from the ranks of these newly trained special service troops. They all realised that the losses might be heavy, but they knew that the rewards for Britain made the operations in which they took part more than worthwhile.

The policy of using only Imperial troops was vital. It was based on the fact that as casualties would be high, the unhappiness would be equally shared by the whole of the British Isles and Commonwealth. Virgin soldiers had no part to play in Combined Operations.

After his 1940 appointment by the Prime Minister, Sir Roger Keyes had firmly established Combined Operations Command and, in spite of a shortage of men and equipment and cooperation, achieved some major successes. Despite his age he was still full of energy and had managed to imbue the aggressive spirit into the mainly volunteer force of Commandos.

Leading an organisation of that kind and at that time was a very demanding task. Not only did he have to wheedle the right type of recruits out of the army, navy and air force, but he was also expected to carry out Churchill's orders, which were to establish a vast organisation and build up a specialised fleet with the express object of returning to continental Europe. This all we must remember, within months of our well-established British Expeditionary Force of more than a million men having been forcibly ejected from France.

To have taken on this impossible task gives an idea of the measure of this wonderful man. More to the point, he had been most successful in all the operations undertaken.

Unfortunately, Keyes had fallen foul of the Chiefs of the three services from whom he had to extract the men, material and ships. In fact, Churchill had to invariably referee battles not so much with the enemy, but between the First Lord of the Admiralty and the Army Chief of Staff, who were in constant combat with the crusty old seadog Sir Roger Keyes, who in reality was their senior.

Churchill had sufficient political and military problems in a war that to some, including secretly myself, seemed all but lost. He urgently needed a man with more patience, together with a modicum of diplomatic skill, to manage the embryo Combined Operations Command.

He would also have to be an engineering officer with a thorough knowledge of naval design, communications and radar, as he was going to have to dream up the various different types of landing craft, together with the support vessels that would be needed for combined operations.

These were going to have to be designed and manufactured with one express object in view, that of landing a massive army on a heavily defended foreign beach.

CHAPTER 12
THE ARCHITECT OF COMBINED OPERATIONS COMMAND

This part of the story really begins at the end of the First World War, when a keen young would-be sailor attended a talk by Admiral John Jellicoe, who was describing his predicament during the vital and dangerous First World War naval encounter, the Battle of Jutland. This naval confrontation had taken place when hundreds of warships of the British and German battle fleets faced each other in the North Sea, off the coast of Germany. The admiral described the frightening situation in which he found himself, when sitting on his flagship only able to communicate with the four ships of his fleet with whom his crew had visual contact.

He had no long-range communications whatsoever with the battleships, cruisers, destroyers and the myriads of smaller craft of which his fleet was composed. He could only hope, 'Nelson' style, that every man would do his duty and that the German navy would succumb. He was aware that the Germans were already using radio communication, but of course only by morse code and using primitive spark transmitters.

Nevertheless, Reinhard Scheer, the German admiral, knew exactly where each and every one of his vessels was located and could exert full operational control over their movements. This was a vital battle for England, a battle in which Britain's mastery of the seas would have been forfeit, if the decision went the wrong way.

Fortunately for Britain, the Royal Navy was successful, but young Louis Battenberg decided that if he could do anything about it, the Royal Navy would never again face a major naval battle without top-rate radio communications.

This was certainly a tall order for a young man about to commence his career in the Royal Navy. He had no power or influence, and in fact had just emerged from his position as a 'snottie' or midshipman, by some regarded as the lowest form of life in the navy.

As a midshipman he suffered all the privations of an apprentice in industry. To sleep, he had to sling his hammock in the only place available, which was under an all-night emergency lamp. Also, as a 'snottie' he was held in disregard by the sailors as well as the ship's officers, one of whom he hoped one day to be.

As in industry, if you were prepared to suffer you might one day aspire to the professional class, but only if you could last the course. Also, as in the outside world, he had to choose a specialty and luckily for all of us, he chose wireless or radio as the profession of choice.

Born in 1900, his father Louis Alexander Battenberg, already a career officer in the Royal Navy, placed him in the Royal Naval College at Osborne in the Isle of Wight. Though just 13 years old, he was already a keen radio enthusiast. He realised more than most that in this infant science he would surely be able to make his mark. Though only a cadet at the college, he had already accumulated much of the limited technology known on this subject and studied the information that was available on the new-fangled science of wireless communication.

Late in 1914, and at the age of 14, he had passed out from the Royal Naval College Osborne with a glowing report stating that, in the principal's opinion, young Prince Louis would make a very satisfactory engineer indeed.

He enjoyed radio technology, and this is why he made such a huge success in this field as his life unfolded. Having qualified, he secured entry to the Royal Naval College at Dartmouth in South Devon. It was here that he showed his worth in technical leadership, coming top of his course in the finals at Keyham (the engineering college, Keyham, Plymouth).

In 1917, when anti-German phobia seemed to have reached a peak, the Battenberg family were forced to anglicise their name to Mountbatten. When he finally received his commission in 1918, he was known as Sub Lieutenant The Lord Louis Mountbatten.

He had survived a terrible blow when his father, who had served Britain well since the previous century, was dismissed from his post as Lord of the Admiralty, simply because his German name made him suspect. This of course, in the middle of the bloody war against the Kaiser.

Such was the measure of the man, because rather than bow out of the navy, defeated by the horrific blow his father had so unfairly suffered, he vowed that he also would one day be First Lord of The Admiralty. This in fact was to happen, and the man who had promoted his father to the post just before the outbreak of the First World War was many years later also to promote his son, Lord Louis. The man was of course, Sir Winston Churchill.

Mountbatten's five years at war in a very junior capacity during the First World War was a superb naval apprenticeship. This was to serve him well when fate decreed that he mastermind and initiate the organisation that was to precipitate the greatest naval operation of all time, the D-Day landings in Normandy, known as Operation Neptune, the prelude to Operation Overlord.

Now, as a young naval officer, Mountbatten spent much of his time thinking and applying his ideas for an ever more impregnable Royal Navy. Of all the services, the navy was the most conservative. Change was anathema to Their Lords of the Admiralty, and Mountbatten in his efforts to secure technical change found himself constantly in the 'dogbox'. This in no way deterred him, and he was to bring about some vital changes in various aspects of naval armament, operations and procedure; but more particularly with respect to communications and the amphibian technology that were to presage victory for the Allied forces in the Second World War.

In 1924, when Mountbatten made his choice of a specialist career, he was posted to the Royal Naval Signal School at Portsmouth. On completion of this demanding and difficult study of the whole gamut of naval radio communications, he secured entry to the Higher Wireless Course at Greenwich.

His years of professional operational experience allowed him, on completing the Greenwich studies, to be admitted as a full Member of the Institution of Electrical Engineers. He was at last a professional specialist and in a position to make his mark on the Royal Navy.

In 1926 he was appointed to the position of Reserve Fleet Communications Officer, where he could now put his ideas into practice. A year later he was promoted to Assistant Fleet Wireless Officer to the Mediterranean Fleet, the reward for coming top of his course, where incidentally he was more a teacher than a student.

Radio was still emerging from the Stone Age and Mountbatten was in his element. He was keen, enthusiastic and worked all the hours of day and night improving his knowledge of the new science, most of which was as yet undocumented. One thing was certain: he intended to be a radio specialist with a complete knowledge of the new science and its future development.

As the development of radio technology was rapidly advancing, Mountbatten decided in 1929 to return to the Signals School at Portsmouth where, as Senior Instructor, he could find time to experiment with the new components that had arrived. Lecturing and teaching a subject helps in no small way to clarify the science, even in your own mind.

New components required new drawings of the new devices and Mountbatten tackled the problem by setting up his own drawing office from which emerged a long series of standardised symbols for all the new components, many of which are in service to this very day.

This came about when he took the classic standard radio instruction manuals, known as the 'Admiralty Handbooks', and redrafted all the drawings in such manner that they could be more easily understood. His clear-minded approach to the as yet embryo science helped many students, including the writer himself, to master the new science of radio communication.

By standardising the designation of components and simplifying much of the instructional text, he earned the undying thanks of thousands of radio students. Needless to say, both volumes of the *Admiralty Handbooks on Wireless Telegraphy* became the standard work prescribed for use in technical colleges all round the world.

In the mid-thirties, Mountbatten was promoted Chief Radio Officer to the Mediterranean Fleet at the time when Mussolini had attacked Libya and Abyssinia.

Finding communications security and discipline very slipshod, he set up a new code of communication ethics and personally policed the radio bands, setting up a very advanced, highly disciplined system of radio traffic control.

With the threat of war with Italy, the Admiralty ordered most of the Mediterranean Fleet out of Malta to a safer location in Alexandria, Egypt. Only a token force, Mountbatten's First Destroyer Flotilla was left in Malta, as an attack from the massive Italian fleet and the Italian air force seemed a real possibility.

Mountbatten laid on a series of exercises in which for the first time, the three services were controlled by a specially prepared Communications Headquarters Ship, HMS *Wishart*, which would lead his tiny defending flotilla when it put to sea.

For the first time ever, he could exert radio control not only on the navy, army and air force units, but also on his submarines. This was in fact a revolution in military science and the concept that was to lead to the development of the Headquarters (HQ) ships, HMS *Bulolo*, *Hilary* and *Largs*, the ships that were to control the miracle of the D-Day landings.

Churchill, always the naval person, noted the resounding success of the Malta exercise and when, many years later in time of war, a younger leader with a wide-ranging knowledge of engineering and communications was needed for Combined Operations Command, he knew just where to look.

After Malta, young Mountbatten was at long last offered the command of a destroyer, HMS *Kelly*. As the ship was still under construction, he pleaded for permission to build a more battleworthy bridge from which to control his ship more effectively under battle conditions.

Initially the Admiralty turned down his request, but after consideration, his criticisms were grudgingly admitted to be well-founded. They allowed Mountbatten to have his way and furthermore permitted the construction and installation of his other inventions that would materially assist a destroyer commander to control his ship with more precision under active service conditions.

Furthermore, on inspecting the completed HMS *Kelly*, they ordered that all new destroyers now be built to the new design.

At the same time, dissatisfied with the hitting power of the destroyer's tiny defensive armament, the First World War Vickers machine gun, in 1937 Mountbatten approached a Swiss engineer who had developed a cannon with an extremely high rate of fire. The gun could fire light armour piercing explosive shells rather than 0.303 bullets and Mountbatten realised that the weapon would be an ideal replacement for the Vickers machine gun.

His constant quest for improvements in naval vessel design caused much resentment back at the Admiralty in Whitehall. When it was discovered that he had approached a commercial company with the request for the supply of this new high-speed cannon, it seemed that he had at long last overstepped the mark.

Nevertheless, Antoine Gazda brought his magnificent new gun to England, where it was demonstrated to be an excellent and wonderful weapon.

As ever, the Admiralty automatically turned it down. Two years later, when Britain was already at war, the Royal Navy frantically ordered from neutral Sweden 400 units for immediate delivery. The story has it that we had to wait in line until a German order had been completed.

In August 1942 it was a barrage of these guns, fired by the destroyer HMS *Albrighton*, which demolished a German gun position that was enfilading Green Beach, Pourville during the Dieppe Raid. This thankfully allowed my escape together with about 250 men of the South

Saskatchewan Regiment and Queen's Own Cameron Highlanders of Canada, who had been trapped on the beach and who were about to become prisoners of war.

As a communications officer, Mountbatten was of course very concerned about secret naval communications if war was declared. It was common knowledge that the German navy during the First World War had a great advantage in that they had broken our communications codes.

In 1937, having seen the RAF Typex coding machine, a wheel-based electro-mechanical cipher machine, an adaptation of the commercial German Enigma, in action at Headquarters Fighter Command, RAF Station Stanmore, Mountbatten pleaded with the Admiralty to adopt the same or a like system for naval use as war seemed imminent.

As ever, his suggestion was turned down, but once again adopted in great haste once war had been declared.

Heavily involved in electronics, Mountbatten knew all the detail on the RAF secret weapon: radar. He wanted a unit for use on his own ship but as yet, no British naval type of radar had been developed.

In typical style he purloined an ASV I radar from a grounded Fleet Air Arm Swordfish aircraft and installed it himself aboard his own ship. The ASV I was a low-power airborne anti surface vessel radar, but he could at least see surface vessels over the horizon and under any weather conditions. He could tell whether a ship was to the port or starboard and its range in miles. On a large ship he could expect a range of at least 10 miles by day or night. At this point in time, this was a stupendous step forward.

Belatedly, only in fact in 1940, the RAF ASV II, which plays an important part in this saga and now under the designation of Naval Type 286, was fitted to the masthead of all destroyers.

War broke out in September 1939, and HMS *Kelly* was almost immediately involved in action off the coast of Norway.

When his ship was torpedoed and about to sink, Mountbatten off-loaded his crew onto an attendant destroyer, arranged a tow and brought the crippled *Kelly* back to port.

A year or so later, when Crete was under attack by the German army supported by the mighty Luftwaffe, Mountbatten went to the assistance of the land-based troops of General Freyburg VC in an area where two cruisers and destroyers had just been sunk. For a while he helped the hard-pressed troops by bombarding the targets requested, but in short measure the *Kelly* itself was sunk by German dive bombers. Mountbatten had to swim for it, but was picked up and thankfully avoided capture by the victorious Germans.

In 1941 he was overjoyed when the navy now offered him the command of the aircraft carrier *Illustrious*, which had been grievously damaged in the Mediterranean fighting. It was being repaired in the shipyard at Norfolk, Virginia. He was in the United States and had just managed to get his shipmate survivors posted to his new command, when Churchill ordered him back to England for a new and very secret assignment.

Mountbatten was very disappointed. He came back to London, where Churchill informed him that he was to replace crusty old Admiral Sir Roger Keyes and become Technical Advisor of the embryo Combined Operations Command.

Sir Roger Keyes had already made very considerable progress in the setting up of Combined Operations and had successfully carried out raids on northern Norway. As mentioned, being of approximately the same seniority as the Chief of Naval Staff, he found great difficulty in securing cooperation from officers who regarded him as a potential threat. Furthermore, the Admiralty did not want to relinquish control of future operations that involved beach landings.

Combined Operations Raids 1941/42

Operation Claymore	Lofoten islands, Norway	03/1941
Operation Gauntlet	Spitzbergen, Norway	08/1941
Operation Anklet	Lofoten Islands, Norway	12/1941
Operation Archery	Vaagso, Norway	12/1941
Operation Biting	Bruneval, France	02/1942
Operation Chariot	St Nazaire, France	03/1942
Operation Jubilee,	Dieppe, France	08/1942
Operation Musketoon	Glomfjord, Norway	09/1942

Keyes had a mandate to carry out Commando raids on German occupied territory, but as previously noted, had been forced to spend much of his time in combat with the Chiefs of Staff.

Frustrated, Churchill wanted a man with diplomatic skills who would be able to sit at the same table with the Chiefs of Staff and not be in permanent combat with them. He needed a man with an engineering and communications background to transform the tiny Combined Operations Command, which then had about four Glen Line Troop Carriers and a few dozen vintage so-called beach landing craft, into a gigantic invasion armada.

Churchill realised that time was at a premium and wanted no argument on the magnitude of the problem. He needed an individual who would ride roughshod over all the negative thinkers and build up a specialised invasion fleet and train the seamen to man the new ungainly flat-bottomed craft.

Hopefully, young Mountbatten would be able to accomplish this task without the constant in-fighting.

The net result of this bickering had been the constant cancellation of proposed actions after long discussion. This infuriated Churchill, who was now at the end of his tether. Though he had to forego command of HMS *Illustrious,* a state-of-the-art radar-equipped aircraft carrier, Mountbatten, with Churchill's prodding, set to work with great gusto to carry out the Prime Minister's instructions.

Initially his assignment was to keep the German troops occupying France on tenterhooks by carrying out Commando raids. After their raids they would leave behind the grisly bloody remains of German soldiers killed during their surprise attacks, usually carried out in the dead of night. Their activities kept the coastal German occupation troops scared and terrified. Hitler was not amused, as the very word Commando made the young Germans shudder.

They were a rough seasoned army whom the German troops occupying France learned to fear and hate. Most of these bitter fighting men had lost relations and friends in the Dunkirk disaster and many were motivated by revenge. Losses, danger or risk were not in their vocabulary.

CHAPTER 13
BRITAIN'S SALVATION

By the end of 1941, radar, or RDF as it was then known, had proved itself to be a wonderful defensive weapon.

More than a year previously we had staved off Operation Sealion, Hitler's invasion of Britain, by the skilful use of radar. The chain of CH and CHL radars had effectively multiplied our tiny fighter force by allowing them to be at the right place and at the right time without wastage of engine hours and terribly expensive standing patrols.

These two radars, the CH (the brainchild of Arnold Wilkins), and the CHL (really an offshoot of W.A.S. Butement's radar CDU (Control Display Unit)), were our only two surveillance radar systems. The CH could read heights accurately enough for the control of day fighter operations out to a range of 100 miles and more, and the CHL backed up the CH by detecting any aircraft which flew below 3,000 feet.

The other wonderful invention that made the task of the German bombers even more difficult was IFF, the Identification Friend or Foe unit. This was a 14-inch cube, grey-painted aluminium box mounted behind the pilot in the cockpit of the Hurricane and Spitfire aircraft. The IFF unit gave the CH station a clear definite signal from a friendly aircraft, allowing hostile aircraft and formations to be easily identified and attacked.

These three radar devices performed magnificently, and Fighter Command Headquarters at Bentley Priory, Stanmore, received a complete picture of the air situation on all the approaches to the British Isles.

Almost immediately after the Dunkirk period, Reichsmarschall Göring had launched the gigantic air attack on Great Britain. The incredible heroism and devotion to duty of the fighter pilots, backed by the smoothly oiled machine of Fighter Command, prevented the Luftwaffe from gaining the victory, which had to precede Operation Sealion, the invasion of Britain. As far as we were concerned, the proposed invasion of England would only now take place in spring 1941.

The next phase was the vicious night blitz that followed the day Battle of Britain. It started from October 1940 when Göring and his Luftwaffe attempted to terrorise the women and children of Britain by bombing and burning their homes to the ground.

There was no way of accurately aiming the thousands of tons of bombs on London, Coventry, Birmingham and many other target provincial cities. With raids like the one on the City of London, in which the German air force dropped 250,000 firebombs in one square mile of the Barbican in little more than an hour, it seemed that we were doomed. With the tens of thousands of casualties that resulted from these terror attacks on the people of Britain, the Germans hoped to force Churchill and the British government to sue for peace. It had worked in Spain, Poland and Holland and now it was Britain's turn.

In noisy tirades to huge, assembled audiences, Hitler threatened to destroy England if we did not immediately acquiesce to his demands. The destruction and casualties mounted alarmingly, as the Luftwaffe discovered that it had nothing to fear from RAF radar at night.

The anti-aircraft gun laying radar (GL) was not effective and the German bomber fleets had it all their own way. At times they motored across London at ridiculously low altitudes, and I distinctly remember seeing the black crosses and reading the insignias on German aircraft caught in the searchlight glare. The Heinkels, Dorniers and Junkers aircraft roared low over London and other cities in the sure knowledge that the RAF was powerless to interfere with their murderous attacks on the working people of Britain.

The battering was compounded by the regular broadcasts of an Englishman, Alfred Joyce, 'Lord Haw Haw'. He broadcast daily to the people of specific towns warning them that they were about to be bombed. As the people of Britain tried to cope with the horrific casualties and fiery destruction, he gleefully gloated over their pain and suffering, at times even quoting the names of the victims.

The raids on Liverpool, Manchester, Glasgow and the other provincial towns were not directed at specific targets. They were directed at the man in the street, and they were succeeding, as the morale of the people had dropped to an alarmingly low level. For women and children to sit all night in a cold wet shelter and yet try to get on with their lives the next day was more than difficult. Especially when so many surfaced to find that their homes had been destroyed.

With huge successes such as the murderous bombing of Coventry, the Luftwaffe was at long last beginning to recover from its defeat in the daytime Battle of Britain.

For eight long months the people of Britain were subjected to an ordeal by fire and explosive and there was no apparent defence.

The writing was on the wall.

It seemed that this overpowering night blitz would be followed by an invasion of Britain and subjugation to the Nazis, who had already drawn up detailed plans for the government of the newly conquered territory. Even the gauleiters for various key areas had been chosen by Alfred Joyce from amongst the ranks of members of his organisation, the British Union of Fascists.

To many of us, though we would not admit the fact, it seemed that there could be only one possible conclusion to the impossibly heavy battering to which we were being subjected. After all, the Germans had already conquered mighty France, Norway and all the Low Countries. Their conquests now stretched the full length of the European mainland, from Norway in the north to Spain on the Atlantic seaboard.

The situation looked more than bleak.

In Spring 1941, the invasion that was expected to follow the harrowing night bombing blitz that caused so much horror, death and destruction to the people of Britain did not materialise. The raids had reached a crescendo in the first week of May 1941 when, quite suddenly, there was a let up in the bombing.

The people of Britain were relieved but uncertain about what Hitler had next in store for them. Was it the dreaded invasion? The citizens of Liverpool, London, Coventry and many other cities and towns licked their wounds and tried to clear up the awful trail of death and destruction left by the merciless terror bombing of the Luftwaffe.

On 22 June came the unbelievable news that the German army had invaded Russia.

Did that let us off the hook? It did not seem so, as the massive German assault, Operation Barbarossa, rolled at high speed across the Russian plains. The Wehrmacht once again was proving invincible. Hitler boasted that the war would be over by Christmas 1941, and then he would be free to deal with Britain. The prospect was not pleasant.

The German battleships roamed the oceans of the world, sinking merchant navy shipping, Allied battleships and destroying convoys seemingly at will.

The U-boats were sinking on average about 700,000 tons of merchant shipping a month and it seemed that they alone were going to bring victory to the Third Reich.

The food ration queues in the cities and towns grew longer and Churchill admitted some years later that nothing scared him more during the course of the war than the U-boat.

Every night merchant navy seamen, with their ships torpedoed, drowned in seas of burning oil, and Britain's food rations grew more and more meagre. True, the Mark II ASV radar was making life far more difficult for the intrepid U-boat men, but they had now found a counter to the equipment and were fighting back most successfully.

The news from Russia worsened every day and by autumn 1941 it seemed apparent that the war would soon be terminated, and Hitler could once again look in our direction.

In the Middle East there was also a catalogue of disasters that made it seem more than possible that our short cut to the Far East and Australia, the Suez Canal, would soon be dominated by the Axis powers.

Roosevelt, who aided Britain in every way possible way, was always on uncertain ground as there were many in the United States, apart from Nazi sympathisers, who would have loved to see the demise of Britain and its Empire, the Empire on which it boasted 'the sun never sets'.

As the year 1941 drew to a close, the people of Britain viewed the future with dismay. Churchill made speeches intended to raise morale, but there was no light on the horizon. Then overnight, the situation changed. The Japanese attacked Pearl Harbor.

America was now our open ally.

CHAPTER 14
FORGING THE WEAPONS

It was on 7 December 1941, the 'Day of Infamy', that the Japanese destroyed a large portion of the US navy at Pearl Harbor.

On 10 December, the battleships HMS *Repulse* and HMS *Prince of Wales*, together with many other naval boats, had been sunk by the Japanese air force, with the loss of 842 men.

Despite these utter disasters, coupled with the uncertainty of our survival in the war against Germany, Churchill decreed that the new year, 1942, would be the year that the Allies, for the first time, would go on the attack.

They were brave words in a new year that was going to be packed with catastrophes in the Middle East, the Far East, Europe, Russia and at sea. Perhaps it was just as well that we could not divine the future, but at long last we had a powerful open ally in the great United States.

It is a great wonder how Roosevelt survived when surrounded by vocal isolationists and many famous personages who were Nazi supporters. With his covert aid to Britain, he was virtually defying US law and might at any time have been exposed to impeachment by his peers.

In spring 1942, we lost our great naval base in Singapore, which fell to the Japanese. India was preparing to welcome the heroic conquerors with their New World Order that was at long last to dispel the white faces of the Raj, and, under the new order, Asia would now be for the Asians.

In Russia the position was just as grave, as the German soldiers were using the tramcars in one remote Moscow suburb. German artillery was hammering Leningrad in the north, and, after massive tank battles, other German armies were racing south to the Caucasian oilfields.

In the Middle East Rommel was rolling back Claude Auchinleck's forces and preparing for Mussolini's triumphal entry into Cairo, mounted on a white charger. Germany and Italy would meet up with a victorious Japan in the Middle East once India had been defeated and the Axis, which would now run on Arab oil, would be consolidated.

In London, mass meetings held by Lord Beaverbrook and other celebrities, including Sir Stafford Cripps, were calling for Churchill to resign. The meetings held in Trafalgar Square, which is located on the northern entry to Whitehall, usually ended in a mass march past the Cenotaph, with the crowd calling for Churchill's blood. The socialist politicians were accusing the Prime Minister of playing off Russia against the Germans. They insisted that by playing politics, he was losing the war.

Their banners read 'Second Front Now'.

It seems incredible now, but the crowds with their passions aroused by self-seeking politicians did not seem to realise that at that time we did not have the ships, the men or the arms to launch a sizeable attack on Hitler's 'Fortress Europe'.

In February, General George C. Marshall was despatched to Britain by Roosevelt to help build a force for Operation Sledgehammer, an attack on France in 1942. At the end of May, with local politicians demanding action, hard-pressed Churchill had to receive a small group of top US generals who had been detailed by the President to put some pep into Britain's war effort. They had taken up residence in Grosvenor Square in the West End of London, from where they made it known that they were not satisfied with the way Britain was running the war.

Led by a disgruntled General Dwight Eisenhower, they accused the War Cabinet of dragging its feet and not putting enough effort into confronting the German forces in occupied France and giving them a sound thrashing. Eisenhower had set his heart on Operation Sledgehammer being launched in late summer, with landing craft yet to be built in the United States. A series of other sacrificial plans had been suggested by the gung-ho American visitors.

The President was pressing Churchill to launch a sacrificial attack on the mainland of Europe in the hope of taking some of the pressure off the hard-pressed Russian army, which seemed to be on the verge of defeat. A tiny force of a few thousand men, which was all that would be available for an attack at that time, would have had precious little effect on the millions of German troops occupying France.

In retrospect the idiocy of the proposal does not make it worthy of discussion, but Churchill, who was depending so much on American assistance, could not simply fob off the President's gentlemanly request. Another definition of a 'gentlemanly request' is an order, but Churchill, with vivid memories of his involvement in the First World

War Dardanelles Campaign, paid lip service to Roosevelt. When, in 1915, Winston Churchill was First Lord of the Admiralty, he had spearheaded the naval attack on the Dardanelles Strait in Turkey, aiming to knock the Ottoman Empire out of the war. The naval attack and subsequent land battle in Gallipoli failed, with heavy Allied losses. This led to his political downfall and his resignation from the government.

At the same time, Mountbatten's Combined Operations Command, under a massive workload, gradually got its house in order. Its tasks were wide-ranging, from setting up dozens of training establishments in places as remote as Shallufa, on the Bitter Lakes in Egypt, and Inverary in Scotland. In addition, apart from initial designs on a multitude of beach landing craft, which would all have to be somehow rigorously assessed before mass production could be commenced, there was the requirement to obtain adequate suitable personnel from all the armed services and train them in the vagaries of beach landing operations.

There was so much to do and little time to accomplish all the different tasks that it was hoped would allow the Allies to make a successful attack on mainland Europe one day.

With more than 2 million German troops in France, at least 500,000 Allied troops would have to be put ashore if we were going to stand a chance of defeating the well-established German garrison forces.

Obviously, they would not all be able to land at the same time, but to land just 100,000 troops would take some thousands of specialised landing ships, which still had to be designed and built. Certainly, much of the material could be transported in freighters, but only after a measure of security had been established at sea and in the air.

At the beginning of 1942 the shipyards of Britain were frantically building cargo vessels to make up for the incredible losses due to U-boat activity. The shipyards were trying to build merchant ships to replace the 700,000 tons, on average, of shipping being sunk every month by the U-boats. The various types of flat-bottomed beach landing craft that might be needed for an eventual invasion had a low priority.

In 1941 the British Army, with trucks and cars disguised as tanks, had routed the Italian armies in North Africa. Having advanced hundreds of miles across the Western Desert, this comical force, with little effective fighting power, was suddenly confronted with the requirement to consolidate the territory gained.

The problem immediately took on a new dimension when the German Afrika Korps landed on the North African coast. Before the next round of fighting started, the Allied generals pleaded for support, supplies, petrol, water and ammunition. The desert roads along the North African coast were merely tracks, which were rough, rocky and dangerous. High speed movement of supplies to support the armies, now out on a limb, was not possible for many reasons. Also, there were practically no trucks apart from those captured from the Italians.

As the Germans prepared to attack, there was the urgent requirement to move all the necessary military paraphernalia west from Egypt to give the Allied troops a fighting chance. It was decided to use beach landing 'lighters' (later named Landing Craft) to keep the armies supplied. These were flat-bottomed craft with a large ramp down which trucks could debark directly on to a beach. These craft were built in British shipyards and shipped in four sections, which were assembled in Port Fouad on the east bank of the Suez Canal, with the town of Port Said on the west.

Whilst the army in the Western Desert was preparing to go into combat with Field Marshal Erwin Rommel's Afrika Korps, the men working on the assembly of the lighters spent their leisure hours sunning themselves on the beach of the French Club in Port Fouad. This made their job more than bearable.

Most of the freighters carrying the lighter sections had to make the long trip round the Cape of Good Hope and through the Red Sea, as passage through the western Mediterranean was impossibly hazardous for the slow merchant vessels. With time at a premium, the lighters had to be hastily assembled and thrown into service immediately, as need was great.

Because of the hasty design and construction, many of the mistakes had to be rectified by the crew whilst in service. There was still so much to be learned about the beach landing craft, and this of course fell on the shoulders of Combined Operations Command, who listened and rectified design defects before the next units were constructed and despatched.

Some of the faults were hair raising, as was the case with the first units that were supplied with petrol engines that seemed to have been obtained from a junk yard. The main problem was that the multiple screws that propelled the craft all turned in the same direction. The flat-bottomed craft were difficult enough to steer as they had no keel, and it was found almost impossible at times to steer to the starboard as the screws gave a permanent bias to the port.

After blistering comments were carried back to Combined Operations HQ in Whitehall, diesel engines were fitted on to these tough workhorses, and they were renamed Tank Landing Craft or TLC. Later, when the United States renamed the different Combined

Operations landing craft, they became the famous LCT, which did trojan work in all the theatres of war.

With heavy losses in men and materiel taking place all around the world in every theatre of war, the building up of Combined Operations Command was a thankless and difficult task. The many branches and activities of Mountbatten's command almost defy description. Its purpose, as detailed by Churchill, was to train and build up a force for raiding occupied territory with the ultimate objective of placing a force ashore on mainland Europe that could meet and defeat the long-established German army. With the crushing defeat at Dunkirk still fresh in our memories, this was indeed a tall order.

Mountbatten's new directive was issued on 16 October 1941. His responsibilities were defined under four heads:

'Under the direction of the Chiefs of Staff you will:-

1. Act as technical adviser on all aspects of, and at all stages in, the planning and training for combined operations.
2. Be responsible for co-ordinating the general training policy for combined operations for the three Services. You will command the Combined Training Centres and Schools of Instruction.
3. Study tactical and technical developments in all forms of combined operations varying from small raids to a full-scale invasion of the Continent.
4. Direct and press forward research and development in all forms of technical equipment and special craft peculiar to combined operations.'

Part of Mountbatten's task as Chief of Combined Operations Command was to formulate and oversee the basic design for the dozens of various types of beach landing and support craft. A major problem was to inveigle men from the other services and the Ministry of Labour, and then train the tens of thousands of them to man the strange new craft that Combined Operations had in mind.

It was pretty obvious that eventually, perhaps 5,000 craft would be needed for the initial assault on Europe, and that thousands of service vessels would be required to build up and maintain a steady flow of all the necessities of life for the army, navy and air force once they had taken up residence.

This situation would be particularly frightening and dangerous during the first month of the landing. These were, of course, only some of the prerequisites.

To get Mountbatten's apparently impossible task into perspective, it was going to be like moving a large industrial town into France with all the facilities to maintain its citizens on a daily basis, and this to be carried out on a foreign shore with a German occupying force of more than 2 million soldiers guarded by some 500 assorted radars and thousands of coastal defence guns, many radar directed.

Apart from this project, the eventual invasion of Europe, was the requirement to meet any panic tactical requirements of the War Cabinet, such as the capture of Madagascar (Operation Ironclad), far away in the Indian Ocean, before the Japanese took up occupation. They would certainly turn it into a U-boat base on our highway to the Far East and, for that matter, via the Red Sea, the Middle East.

This Combined Operation mounted by Britain was to capture an island with a hostile French garrison on the other side of the world. It went off like a charm and the fleet returned to England in good time for the next operation, in Spring 1942.

Another vital and gigantic operation that Roosevelt regarded as urgent was in the process of being planned at Combined Operations Headquarters. This was the ousting of the Vichy French forces and the occupation of the whole northwestern corner of North Africa. This major operation that superseded all others included the landing of whole American armies, which would embark from US ports, some landing on the tidal Atlantic coast of French Morocco and others in the comparatively passive Mediterranean ports of Oran and Algiers.

These were undoubtedly vital strategic dispositions that were to take place ostensibly in Autumn 1942.

Though almost frivolous when compared to the magnitude of the proposed North Africa landings, American generals now in residence in London were demanding that Combined Operations carry out a raid on Nazi-occupied France. It was to be known as Operation Sledgehammer. It took no genius to see how dangerous and impossible this would be in Spring 1942, as we did not have the men, arms or even the ships to lift a large force into France at that time. General Eisenhower and his cohorts were still accusing the British of dragging their feet and not being prepared to confront the Germans in open battle. When finally told that this operation could not be implemented, Eisenhower threw a tantrum and threatened to pack his bags and go back to the United States. At that time, it seemed all so simple to him, but events were to force him to grudgingly and belatedly revise his opinion.

Mountbatten had no illusions about the immensity of the task he had been given. With major problems to be solved on equipment needed and strategies to be devised, he had little time to pander to the frustrations of 'gung-ho' soldiers who wanted to see immediate land action in occupied France.

Yet Churchill, not wishing to be seen by Roosevelt as negative and uncooperative with the US army, asked Mountbatten to give the matter serious thought. Eisenhower insisted vociferously that 'Sledgehammer' was urgently necessary to relieve pressure on the Russian army.

The immense buildup of defensive equipment, radar and guns by the Wehrmacht, however, seemed to preclude any major attack on the coast of France. With the horrors of Dunkirk still fresh in British minds, the prospects of ever being able to place a large army ashore in face of the well-publicised and ever-expanding German defences seemed more than remote.

Churchill had been dismissed from his post as First Lord of the Admiralty during the First World War for launching an invasion of the same ilk in the Dardanelles, which had resulted in unbelievably horrific and unbearable losses of men and materiel. The German build-up of defences and the possibility of another disaster on the same scale, but closer to home, was more than he could stomach.

Presumably for this reason he always preferred to think of an invasion in a more remote area such as in the Balkans or the Mediterranean, where the full strength of German defences would not be encountered.

Continually thinking of men's lives, a remote invasion, where the full weight of our naval power could be used against the limited defences that the Germans could arrange in possibly hostile territory, always seemed to appeal to the old war horse.

The problems of attempting a landing in the heavily defended area of occupied France sounded suspiciously like the preparation of another bloodbath, but this time nearer home. Churchill had no illusions about the magnitude of the task he had placed on Mountbatten's shoulders, and he realised that all Mountbatten's efforts should be concentrated on building up Combined Operations Command.

Combined Operations depots and training establishments were set up in India, Egypt, Malta and, of course, all over the British Isles. That this enormous organisation grew so quickly was a wonderful achievement on Mountbatten's part. He was a great salesman and achieved much by sheer persuasive power. The men revered him because he spoke their language and the spirit he engendered made every man a leader. He knew only too well that one day they would all be put to the test.

A major problem at all times was to secure suitable motivated men to take on the very dangerous work as, for the raids, only the toughest of the tough could be used.

These were the volunteers, whom I have referred to earlier, who had joined the Commando units and, together with the airborne troops of various kinds backed by the men of the Royal Marines, they gave Mountbatten a fighting force upon which he could rely for even the most bloody of operations.

If the noisy politico military pundits could be held at bay, he would be able to get on with the real job; that of building up a gigantic, specialised invasion force the like of which the world had never seen.

In 1942, just eighteen months after the huge well-established Allied armies had been summarily ejected from mainland Europe by the victorious German army, Mountbatten and his Combined Operations Command had been ordered by the War Cabinet to prepare the men and equipment for a return to the Continent. At that point in time, this was an all but impossible task.

Hitler had poured millions of tons of concrete in setting up what he proudly called 'Festung Europa', Fortress Europe. This consisted of a number of self-contained mini-fortresses known as *Stützpunkte*. Each was based around huge guns, usually in groups of four, sometimes more; 150-mm artillery pieces with a useful range of 15 miles or more. What made these gun batteries particularly dangerous was that most were radar controlled, which the Royal Navy had learned to its cost in small coastal raids on occupied France.

Each gun battery was protected by a flanking small artillery unit, which in turn was defended by carefully positioned machine guns. These in turn were protected by a sizeable force of mobile infantry. All the Stützpunkte were covered by flak towers, which could give air defence as well as help to defend the countryside around the strongpoint. Prepared with characteristic German thoroughness, they were going to be hard nuts to crack.

The Combined Operations' task was made even more impossible by the basic requirement for any invasion of the Continent. This was the need to move some thousands of naval craft, many crammed with troops, arms and equipment across the English Channel without detection. One thing was certain, if they were detected, the long-range radar-directed naval guns would open up and destroy the impudent invaders.

At the beginning of 1942 we had already detected about thirty coastal German radar installations, and other stations were being commissioned at an alarming rate. The simple fact was that if we could not defeat the German radar, an invasion of the Continent would never be possible.

A portrait of Jack Nissenthall taken at about the time of his enlistment in 1940.

Jack Nissenthall pictured in later life.

A Ground Controlled Intercept, or GCI, Chain Home Low convoy set-up.

Situated to the east of Arundel in West Sussex, the Chain Home radar station at Poling was one of the first twenty planned before the Second World War. However, it had not been completed when war broke out. This picture of three transmitter towers at Poling was taken from the south side. Poling's receiver towers, which numbered four in total, are just out of the picture to the right. (Historic Military Press)

A view of the receiver and transmitter masts of Swingate radar station at St Margaret's at Cliffe near Dover, Kent. This picture was taken by a German propaganda photographer using a long range lens from the cliff top at Cap Gris-Nez. (Historic Military Press)

The 360ft tall transmitter towers at Bawdsey Chain Home radar station, Suffolk, as they appeared in May 1945. (via Historic Military Press)

A typical German Freya aerial array.

The Würzburg radar station at Bruneval on the French coast. It was reconnaissance photographs of this site, taken in January 1941, that led to plans for Operation *Biting* being drawn up. This raid, an early action against German radar technology and equipment, was carried out on the night of 27/28 February 1942. (Historic Military Press)

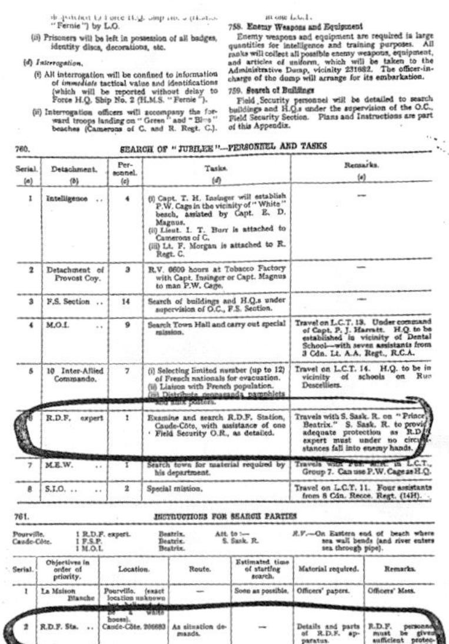

[illegible] "Fernie") by L.O.

(ii) Prisoners will be left in possession of all badges, identity discs, decorations, etc.

(d) *Interrogation.*

(i) All interrogation will be confined to information of *immediate* tactical value and identifications (which will be reported without delay to Force H.Q. Ship No. 2 (H.M.S. "Fernie").

(ii) Interrogation officers will accompany the forward troops landing on "Green" and "Blue" beaches (Camerons of C. and R. Regt. C.).

758. Enemy Weapons and Equipment

Enemy weapons and equipment are required in large quantities for intelligence and training purposes. All ranks will collect all possible enemy weapons, equipment, and articles of uniform, which will be taken to the Administrative Dump, vicinity 231682. The officer-in-charge of the dump will arrange for its embarkation.

759. Search of Buildings

Field Security personnel will be detailed to search buildings and H.Q.s under the supervision of the O.C., Field Security Section. Plans and Instructions are part of this Appendix.

760. SEARCH OF "JUBILEE"—PERSONNEL AND TASKS

Serial. (a)	Detachment. (b)	Personnel. (c)	Tasks. (d)	Remarks. (e)
1	Intelligence ..	4	(i) Capt. T. H. Insinger will establish P.W. Cage in the vicinity of "White" beach, assisted by Capt. E. D. Magnus. (ii) Lieut. I. T. Burr is attached to Camerons of C. (iii) Lt. F. Morgan is attached to R. Regt. C.	—
2	Detachment of Provost Coy.	3	R.V. 0600 hours at Tobacco Factory with Capt. Insinger or Capt. Magnus to man P.W. Cage.	—
3	F.S. Section ..	14	Search of buildings and H.Q.s under supervision of O.C., F.S. Section.	—
4	M.O.I. ..	9	Search Town Hall and carry out special mission.	Travel on L.C.T. 13. Under command of Capt. P. J. Harratt. H.Q. to be established in vicinity of Dental School—with seven assistants from 3 Cdn. Lt. A.A. Regt., R.C.A.
5	10 Inter-Allied Commando.	7	(i) Selecting limited number (up to 12) of French nationals for evacuation. (ii) Liaison with French population. (iii) Distribute propaganda pamphlets and [illegible] posters.	Travel on L.C.T. 14. H.Q. to be in vicinity of schools on Rue Descelliers.
6	R.D.F. expert	1	Examine and search R.D.F. Station, Caude-Côte, with assistance of one Field Security O.R., as detailed.	Travels with S. Sask. R. on "Prince[illegible] Beatrix." S. Sask. R. to provi[illegible] adequate protection as R.D.[illegible] expert must under no circu[illegible]stances fall into enemy hands.
7	M.E.W. ..	1	Search town for material required by his department.	Travels [illegible] in L.C.T., Group 7. Can use P.W. Cage as H.Q.
8	S.I.O.	2	Special mission.	Travel on L.C.T. 11. Four assistants from 8 Cdn. Recce. Regt. (14H).

761. INSTRUCTIONS FOR SEARCH PARTIES

Pourville. Caude-Côte. 1 R.D.F. expert. Beatrix. 1 F.S.P. Beatrix. 1 M.O.I. Beatrix. Att. to:— S. Sask. R. *R.V.*—On Eastern end of beach where sea wall bends (and river enters sea through pipe).

Serial.	Objectives in order of priority.	Location.	Route.	Estimated time of starting search.	Material required.	Remarks.
1	La Maison Blanche	Pourville. (exact location unknown [illegible] white house).	—	Soon as possible.	Officers' papers.	Officers' Mess.
2	R.D.F. Sta. ..	Caude-Côte. 206683	As situation demands.	—	Details and parts of R.D.F. apparatus.	R.D.F. person[illegible] must be give[illegible] sufficient protection to prevent his falling into enemy hands. (F.S.P. and S.O.E. to as[illegible] R.D.F. expert)

Note.—R.D.F. Station sho[illegible] [illegible]hed as soon after capture as possible. On conclusion, party will proce[illegible] Cage via 6 Bde. H.Q.

Orders and instructions for the Dieppe Raid – the circled sections relating to the part that Jack Nissenthall played in the operation.

A cavity magnetron.

Landing craft run in to the beaches at Dieppe under the cover of a smoke screen. This is the kind of scene that Jack Nissenthall would have witnessed on 19 August 1942. (Department of National Defence/Library and Archives Canada; PA-113244)

Landing craft manoeuvring offshore, under cover of a smoke screen, during the Dieppe Raid on 19 August 1942. (Historic Military Press)

The town of Pourville, to the west of Dieppe, as viewed from the eastern headlands. It was on this stretch of beach – Green Beach as it was codenamed – that Jack Nissenthall landed during the Dieppe Raid on 19 August 1942. The photographer is standing on the eastern headlands, in which area the radar station was also sited. (Tommy Larey/Shutterstock)

Part of a Würzburg radar installation on the Normandy coast, pictured after D-Day on 22 June 1944. (NARA)

The surviving Würzburg FuMG.65 radar at Douvres-la-Délivrande in Normandy, France. This was the primary Luftwaffe radar station in the area. (Shutterstock)

Jack, on the left, pictured with Willi Weber.

Jack, in the centre, at NORAD, North Bay.

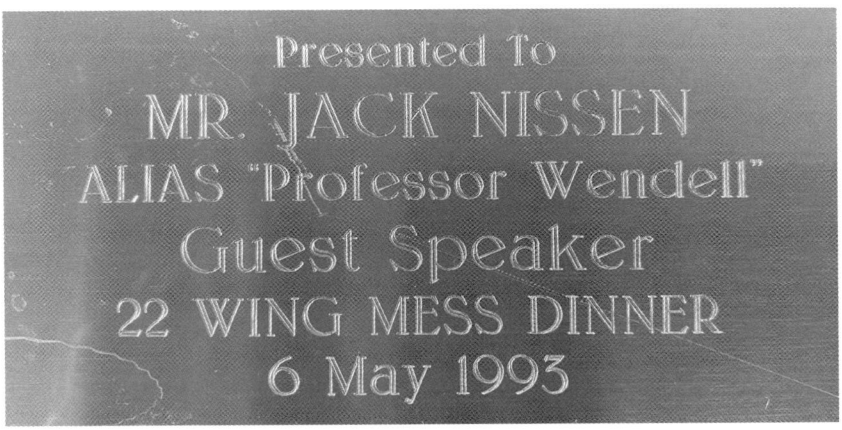

A commemorative plaque presented to Jack during his visit to NORAD.

Furthermore, if we started to jam German radar stations, they would do precisely what we had done in the pre-war years and develop countermeasures for the various jamming signals. We knew from experience that 'Railings' and 'CW' interference could be quickly dealt with by reasonably simple electronics.

Radar pulse jamming at high recurrence rates (50 to 150 kHz) results in an image on a radar indicator resembling fence railing.

Continuous Wave (CW) jamming is a simple yet effective method in which a jammer continuously emits a single frequency or a narrow band of frequencies. This makes it difficult for receivers to process the incoming signals as they are flooded with a steady stream of unmodulated signals.

If we steadily used the 'Mandrel' type of jamming, they would use our signals for experiment and after perhaps six months prepare their own 'AJ' (anti-jamming) tube with optical filters, as we had done in 1939. For this reason, Victor Tait, the Director General of RAF Radar, counselled against using Mandrel on a regular basis, and certainly not before we knew which German radar had to be targeted. The type of countermeasure would, of course, depend on the wavelength of the German system.

Knowing how long it had taken us to develop a counter to Mandrel gave us an idea of the time the German scientists would need to duplicate our effort. With a genius like Manfred Von Ardenne on the job, it would not take very long for them to come up with the answer. Logic dictated that the three radars would have to be tested, and their long-range precision unit definitively identified.

In February 1942, the Bruneval raid showed that the Wurzburg could be scratched off the list. It was of ridiculously low power and in its present form could not have a long-range performance.

The Seetakt was seen to be a naval gunnery radar, which indicated that it would definitely have precision plotting capability, though its range performance would most likely be just over the horizon with perhaps a few extra miles for early warning and pick up of target.

This left only the Freya radar. Our long experience with wide beam CHL type radars suggested that it could not be a precision equipment. This was as long as they had not thought of the PPI (Plan Position Indicator) display, which drew an electronic map of the area around the station and gave our equivalent radar the CHL a fair degree of bearing plotting accuracy.

There was another possibility. A few long wavelength coastal defence fire control radars had been made in England and though they gave a high degree of bearing accuracy at short range, they required a lot of maintenance.

In view of the fact that microwave radars were being developed, the CDU (Control Display Unit) as it was called was to all intents and purposes abandoned.

The other frightening possibility was that the German scientists who were the inventors of the split anode magnetron had taken the extra step and discovered the cavity magnetron, our top-secret microwave transmitter. There was no evidence of this type of narrow beam transmission, but that could have been because we were looking in the wrong places. At that time there was little that we could do about that frightening possibility.

It was quite obvious that the next radar to be investigated would have to be the Freya. Of the thirty units we had now traced, there was a convenient one at Wissant, near Calais, but this was located in a heavily defended security area. There was another station on a headland at Cap de la Hague, near Cherburg, but that was too distant.

Another convenient Freya was at Pourville, just 3 miles west of Dieppe. Combined Operations wasted no time. This vital investigation had to be carried out quickly as we needed to have a yes or no. Had the Freya in some magical way been turned into the main German precision plotting radar for coastal surveillance – or not?

Our information was that it had been improved. If this could be verified, Victor Tait, the Director General of Signals, would have to secretly order the design and manufacture of masses of a suitable type of jamming equipment to work on the 2.5m band. On the other hand, if it was still the imprecise radar we had checked in 1941, the search for the precision radar would have to go on.

In spring 1942, apart from small Commando intelligence gathering raids, under pressure from Churchill, Mountbatten had to address the constant American demand for some larger action that might assist the Russians, who seemed to be on the verge of defeat.

Josef Stalin, the Russian dictator, was also insisting that Churchill's insidious intention was to play the Russians off against the Germans. This, of course, was quite untrue, but hard to refute.

Shortly after his appointment, Mountbatten began to produce results. At the end of 1941 more raids were carried out successfully on the islands off the Norwegian coast. In February this was followed by the raid on the Bruneval radar station and in March the blasting of the docks at St Nazaire in occupied France. These raised public morale no end as at last we were doing something positive against the Germans.

There had been a series of prospective plans for a large landing on the mainland of France, but the one General Eisenhower had been detailed by Roosevelt to initiate was Operation Sledgehammer. This called for the landing of a large force of British troops that would have to draw the Germans into a major conflict. They would then have to extricate themselves and return to England. As we now know, the request was totally unrealistic and amounted to little more than a sacrificial landing. It was really out of the question, because we did not have either the men to carry out such an operation or the ships to carry them to France.

Mountbatten was fighting for more assault craft for his command. The sum total that was at his disposal in early 1942 was twenty-two Assault Landing Craft, about twelve Tank Landing Craft and a small number of tiny unarmoured wooden Eureka craft. There were scarcely enough troops to form even one brigade of the specialised fighting men that he needed for this type of assault.

Despite Montgomery's snide remarks with respect to the advantages of using troops specially trained for the rigours of beach landings, Mountbatten insisted on only using men who had been through his Combined Operations training. As combat was always likely to be at great advantage to the defending Germans after a beach landing, he steadfastly refused to use conventional soldiers. He felt that they could not possibly perform as did, for example, the men of the Royal Marines or the Commandos of Achnacarry.

The fact that there were insufficient ships, arms or specialised equipment available for a major attack did not deter the Americans, the Russians and his political opponents at home from keeping the pressure on Churchill. Operations that were under consideration were raids on Norway, Sledgehammer, Roundup and numerous other schemes, which were presented to the Chiefs of Staff, but each and every one was vetoed for one reason or another, leaving Churchill frustrated and the Americans very annoyed.

They felt that the British were not trying hard enough. In fact, General Eisenhower, when informed that Sledgehammer was not going ahead, stated, 'This would be a black day in Anglo American relations', and that he would return to the United States forthwith. He even suggested that as the British were not prepared to carry out his wishes, the Americans might change their minds and tackle Japan before Germany.

That was never really on the cards, but at that time Eisenhower had no real conception of the dangers of an operation of that type, and limitations imposed on Mountbatten due to lack of ships, equipment and trained men. Meanwhile the reasonable weather time slot during which these seaborne operations might be carried out was slipping by. It should be remembered that there were only about four days in each month where the tides and moon favoured surprise landings. Also, it was mainly the spring and summer months that were suitable for combined operations landings.

CHAPTER 15
BREAKOUT FROM BREST

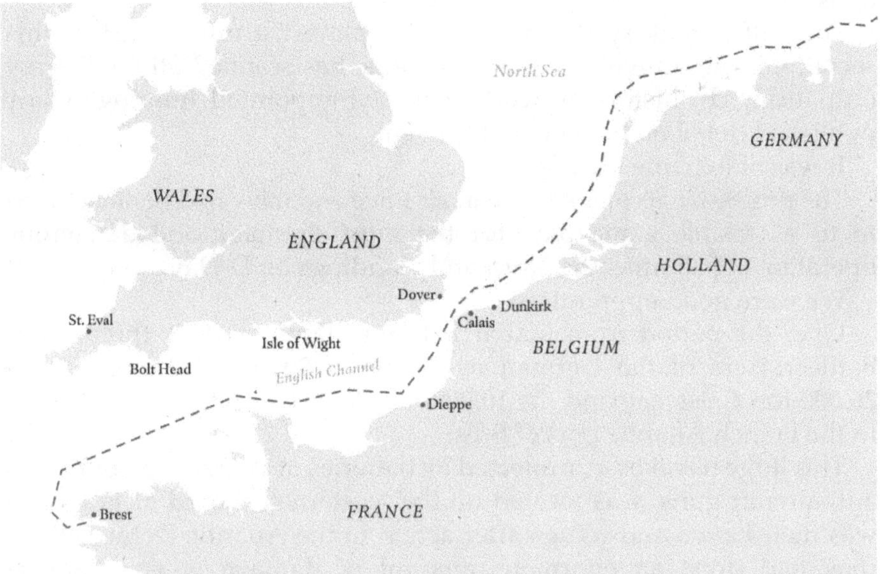

The route of the three German battleships during their escape from Brest, February 1942

These were trying times for Churchill and the War Cabinet, who were under attack from all directions in a war which in early 1942 was going sour in every single theatre of operations.

Winston Churchill wrote of the escape of the German battleships *Scharnhorst, Gneisenau* and *Prinz Eugen* from Brest harbour in February of 1942: '. . . To allay complaints an official inquiry was held which reported the publishable facts. Viewed in the afterlight and in its larger aspects, the episode was highly advantageous to us... but it looked very bad at the time to everyone in the Grand Alliance outside our MOST SECRET CIRCLES.'

They, the Germans, had in fact handed us 'The Key to Victory'.

Even that great German hero and General of Luftwaffe Fighters, Adolf Galland, who together with the Director General of Signals, General Wolfgang Martini, set up the cunning plan for the escape of the battleships, was nonplussed by Churchill's statement.

In Adolf Galland's own memoirs, he, in obvious disbelief, caustically remarks on Churchill's 'amazing conclusion'. This was that in some unspecified and secret way the shocking defeat for the RAF, the Royal Navy and the British army had in some way 'been of great advantage to us'.

For half a century this writer, who played a minor part in this 'secret episode' to which Churchill refers, has scanned all the literary outpourings by historians, worthy and self-appointed, hunting for any published detail on this incredible event.

It was of extreme significance.

The professionals of the German air force and navy had demonstrated to us a feasible game plan for the most dramatic and dangerous operations of all time, the Normandy landings on D-Day.

We were good apprentices.

Over the period from March 1941 into January 1942, three major battlecruisers of the German navy, the 26,000-ton *Scharnhorst*, the 26,000-ton *Gneisenau* and the 10,000-ton *Prinz Eugen* had taken refuge in the French Atlantic port of Brest.

This huge naval base, protected by batteries of Wurzburg-controlled anti-aircraft guns, was located on the western seaboard of France. It was their haven and refuge after action in the Atlantic Ocean, where they had done an enormous amount of damage when attacking shipping on the supply routes to Britain.

They urgently needed a refit if they were to see further action against our convoys, having already destroyed almost a million tons of our merchant navy shipping. Between them they had also sunk the aircraft carrier HMS *Glorious*, the battleship HMS *Hood*, as well as many convoys and their support craft.

The RAF kept a radar eye on the ships by patrolling the harbour mouth and the routes the ships would have to travel in an escape attempt. Long endurance ASV radar-equipped aircraft of Coastal Command waited for them to risk the trip to the German naval bases in the North Sea, where the refit could be carried out. By the end of 1941 RAF Bomber Command had visited them nearly 300 times, losing 45 aircraft and 250 aircrew in the process, but despite much damage, they had survived.

The weeks and months went by with the RAF crews watching the harbour mouth in fair weather and foul, getting more and more bored by the ongoing non-event.

The radar unit used in the aircraft was the simple well-tried original Mark I ASV, which, though very reliable, was not the easiest screen to watch for long periods. The version mainly in use for the surveillance had an electric motor driving a switch on its aerial systems, which was sometimes severely affected by the corrosive sea air, and the radar trace that had to be continually watched was invariably jumping around making the observer's job tiring and difficult.

When the motor required service, the screen showed a kind of vibrating curtain of hash, which resembled what was one day to be called Mandrel jamming.

In the winter of 1941-2, there had as yet been no major jamming of radar by either side, and the RAF observers, not being able to look into the future, were more concerned and annoyed if the split motors, or the 'front end' as they called it, made their long journey of almost 100 miles each way quite useless.

11 February was a cold, dark winter's evening, and the duty Hudson aircraft laboriously lifted itself off the airstrip at St Eval for the routine 'Stopper' patrol.

Leaving the tip of the Cornwall coast behind, the pilot steered due south towards the extreme western cape of Atlantic France, Ushant, to take up his patrol line. Once at altitude, the moonlight lit up the broken cloud and the crew settled down for the usual boring trip.

It was true that the word was out that the big ships were about to make their move, but this story had been spread many times before.

During that very afternoon, the high-flying Spitfire photo reconnaissance aircraft, based at St Eval, had been greeted by the usual anti-aircraft fire and the photographs showed that the protective anti-torpedo booms were still in position around the ships, and there were no signs of the black smoke that would signal that the battlecruisers were 'firing up' in preparation for departure.

When they were 50 miles off the Brittany coast, the radar equipment was switched on and it worked quite normally.

This western tip of France was not a busy operational area, unlike the coast of France 200 miles or so to the east. In that area, Bomber Command aircraft en route to continental targets nightly overflew eastern France, and the Luftwaffe night fighter force, directed by the German equivalent of my own station in South Devon, would try to establish contact.

One or more of the many Freya-type radars would pick up the RAF bombers as soon as they left Britain, and General Galland's ME109 night fighters, fitted with Lichtenstein AI radars, would be guided into an interception as soon as it could be arranged. Both we and the Germans now had this technique off to a fine art and the interception and destruction of enemy bombers for both sides was pretty 'run of the mill'.

However, way out west over the Brittany coast there was very little air activity of any kind at night. That is other than that carried out by my own station, located just a few miles east of Plymouth.

Our activities played out most nights over the English Channel mainly consisted of practice AI interceptions using our own night fighters. Usually, two fighters were involved where one would play fighter and the other target. When the interception had been completed, the two aircraft would change roles. This would usually terminate just before dawn, but it kept the aircraft in position if and when the Luftwaffe were to attempt a raid.

The Lockheed Hudson crews knew that these skies were kept under observation by the Freya radars in and around Brest. For this reason, the observer in the astro hatch always kept a sharp lookout for enemy night fighters, though to date there had never been an interception of a 'stopper' aircraft.

Suddenly, and to his horror and astonishment, the observer spotted a Messerschmitt 110 night fighter that all but collided with the ponderous slow Hudson. In the huge night skies this near accident was more than a million to one chance.

As the power for the ASV radar was taken from the starboard engine, the radar had to be switched off before the pilot could take evasive action. A racing engine would damage the delicate carbon pile voltage regulator unit that supplied power to the radar.

The pilot, with his heart in his mouth, knew only too well that cloud or no cloud the German night fighter using his AI radar could easily locate and overhaul the slow ponderous Hudson and shoot it down.

With its radar switched off, the Hudson dived into the imaginary protection of cloud cover and after a fair period of time, the ASV radar was switched on again.

The set now appeared to be faulty. The screen was saturated with hash. It seemed that the aerial switching motor was playing up. The operator struggled for some time to get a usable picture and finally gave up in disgust. The crew returned to base, as without radar they could do nothing.

After a short time, another ASV equipped aircraft took off to carry on with the surveillance, but on arrival off Brest, it also found its ASV to be unserviceable. This was more than unusual, as the ASV II radar was a remarkably simple and reliable piece of equipment.

Despite all the hundreds of hours of trouble-free surveillance over a nine-month period, on this critical three-hour period the equipment 'broke down'.

As a result, the big ships slipped unobserved past the ASV 'stopper' patrol and sailed first in a northerly direction and then east into the wide expanse of the English Channel.

The German armed services had scored an astonishing victory, defeating the combined efforts of the army, Royal Navy and Royal Air Force, who were backed by the whole of 60 Group, the radar Group of the Royal Air Force. Churchill had to face an irate Parliament and the British public, who were alarmed at the ease with which the combined British defence forces had been summarily beaten.

He was forced to set up a public inquiry and the repercussions of this defeat echoed down the war years.

In his memoirs written after the war, Churchill stated that in our most secret circles, the defeat in some secret way was a great victory.

In fact, as we will see, the Germans had answered our most pressing problem of the time: how on earth were we to assemble thousands of ships off the coast of France in an invasion without being blown to pieces by the hundreds of radar-controlled long-range guns.

By using high power airborne, land-based and shipborne jamming for the first time ever, the German Director General of Signals, Wolfgang Martini had secured the release of the three battle cruisers.

Under the twin secret codenames of Operation Thunderbolt and Cerberus, at Hitler's command the Chief of the Luftwaffe, General Jeschonnek had ordered that a plan be devised to secure the release of the three battleships. His Director General of Signals, Wolfgang Martini had over a period of time secretly manufactured and installed some special equipment to thoroughly jam the 200-megacycle band used by both the Airborne ASV radar and the British CHL ground radar. Also included and playing a vital part in the operation was General of Fighters Adolf Galland.

As the ships were going to steam some 300 miles and negotiate the narrow straits of Dover, air cover had to be supplied over the whole route.

With a limited number of fighters available, Galland used night fighters for that part of the route which would be in darkness and day fighters once the ships reached the area of Dieppe. At high speed the whole flotilla of ships steamed unbeknown to the RAF or Royal Navy into the wide, dark expanse of the western entry to the English

Channel. When passing Cherbourg, they should have been detected by the 'Line SE' patrol, but this did not happen.

For the tired RAF aircrews, the fact that the equipment seemed to have broken down was of little consequence as the Spitfire reconnaissance just a few hours before had shown no sign of the fact that the ships were about to move.

The men all understood that if the battleships were going to sail, they would first have to 'fire up', emitting huge clouds of black smoke for some time, and that this surely would be observed and the appropriate action would be taken by the RAF. Even if the ships left the harbour, they would still have to travel some hundreds of miles through the ever-narrowing sea lane as they approached the Straits of Dover, which at the narrowest point was less than 20 miles wide. That is unless they were brave enough to take the Atlantic route, which would take them past the British Home Fleet based in the Orkney Islands. This was, of course, out of the question.

Taking the Channel route, they would have to run the gauntlet of dozens of British radars that were manned for twenty-four hours a day. This would be followed by the army, with its huge radar-controlled guns at Dover that would most certainly open up on the ships be it night or day.

The chances of their successfully negotiating the English Channel and stepping on England's front doorstep seemed utterly remote. That is how it appeared to Britain's defence forces, ready and waiting. This did not take into account the 'Fog of War' theory, which Mountbatten applied and was amply assisted by the erudite Director General of Luftwaffe Signals, General Wolfgang Martini.

In 1941 I had been posted to a non-existent RAF station, RAF Soar Mill, supposedly located on Bolt Head, the large, bulbous land mass on the southwest coast of England, just to the east of the naval base at Plymouth. The French port of Brest lay some 90 miles to the southwest and Cherbourg some 90 miles to the southeast.

After removing the cows and their evidence from the field, we set up one of the first Ground Control of Interception stations, with the intention of defending Plymouth and Bristol from air attack by night.

We were given a pair of wire mesh runways and two sections of 317 Polish Squadron Spitfires for our protection. Apart from my GCI, which was a night fighter control station, the unit had rapidly turned into a forward satellite station of RAF Station Exeter and from this tiny aerodrome, fighter sweeps called 'Rhubarbs' (search and destroy attacks) were carried out on occupied France.

Squadrons of fighter aircraft visited or rendezvoused over Bolt Head when launching huge wing attacks on occupied France.

Naturally, Luftwaffe aerodromes such as Abbeville and Audembert, where Adolf Galland was known to be based, received much attention during the course of the fighter sweeps. This always resulted in our receiving the unwelcome visitations of Luftwaffe Messerschmitt and Focke-Wolfe fighters on a fairly regular early morning basis. They would approach undetected under the radar beam of CHL Bolt Tail and arrive unannounced, shoot up the Bolt Head installations hoping to find the petrol dumps that sustained our Spitfires, and then make for home with our aircraft in pursuit. Despite the unsuitable weather, the attacks seemed to increase in January 1942, but in February they abruptly stopped.

We, that is the RAF, had been told that Adolf Galland, now a General, had been appointed with orders to go on the attack and that henceforth we could expect trouble. Nevertheless, and strangely, German air activity by day and night had noticeably dropped.

What we did not know was that Hitler had received Admiral Doenitz, General Galland, General Jeschonnek the Chief of the Luftwaffe and General Martini the Director of German Signals at his headquarters in the Wolf's Lair and ordered them to contrive the escape of his battleships holed up in the French port of Brest.

It was a very difficult task and Galland proposed to tackle the problem by setting up a continuous air umbrella of fighter aircraft all along the coast of northern France to allow the battleships imprisoned in Brest harbour to make their way through the English Channel north to naval bases in Kiel and Norwegian waters.

This top-secret order issued by Hitler himself to the air force and navy constituted Operations Thunderbolt and Cerberus.

Galland had secretly arranged a complicated plan for relays of day and night fighter aircraft to provide cover for the battleships on their dangerous trip towards and through the English Channel. The Luftwaffe had a very limited number of operational fighter aircraft available for the massive project. He knew full well that for a large part of the route he would have to contend with the dozens of squadrons of aircraft of 11 Group Fighter Command, for whom he had the greatest respect. Despite the extreme danger both to the battleships and his own career if the project failed, he tackled the almost impossible task in a thorough and professional manner.

In good time General Martini had ordered a number of airborne and ground radar jamming transmitters to be built and installed in night fighter aircraft and at strategic points on the coast of France. The sites chosen were those opposite British radars known to be capable of plotting large surface vessels at long distance. Defeating these radars was an ideal task for Mandrel jammers.

For instance, Dieppe Freya 28, the station visited by the author later that year, had a jammer installed, because it was directly opposite CHL Beachy Head, a radar station known to have an excellent performance based on many previous clashes.

Galland prepared a plan to keep a strong fighter force circling over the battleships at all times as they made along the northern coast of France and past Belgium and Holland on their journey to Kiel. The pilots were briefed to fly very low over and around the ships so as not to arouse the interest of the ever-watchful British radars, which could see aircraft far better than ships.

Towards midnight on the evening of 11 February, Sergeant Ted Cole, the NCO in charge of the watch on my GCI station, had phoned my billet in the Cottage Hotel, Hope Cove complaining of an intermittent fault on the radar equipment.

This was serious as our normal night's work was to keep Beaufighters on patrol over the Channel, ready and waiting for any possible enemy air raids. With no 'business', two Beaufighters would take turn about playing fighter and target. If the equipment were to break down whilst we had aircraft under our control, this would have placed the lives of our night fighter crews in jeopardy as we did all the navigation for the Beaufighters.

On arrival at the station, I could find nothing wrong with the equipment. The controller was in contact with the pilots of the aircraft and the two screens PPI (Plan Position Indicator) and Height Range were functioning perfectly. Ted Cole described what he had seen and insisted vehemently that it was either an intermittent fault on the equipment or truck interference. I pointed out that there were no trucks within miles of the aerodrome, and I asked him to describe what he had seen.

It was very much, he said, as if our operations room vehicle petrol engine had been started up, as the screens were saturated with heavy magneto-type spark interference.

This GCI unit was a mobile convoy with all the equipment mounted in trucks. Once a month for maintenance purposes, the engines of the Crossley vehicles were run up. This was the kind of interference they had observed. I asked the controller what he had seen on his screen.

He told me that whole quadrants of his screen were left brilliantly white for a few seconds, and it was not possible to see the aircraft. I asked him which direction the interference seemed to be coming from, and he suggested that it was from various directions. His assistant, the plotter, however, insisted that it was mainly in the south.

As the interference had only been seen four times and had now ceased, I entered the reports in the transmitter logbook and wearily made my way down to the hotel.

Mid-morning on 12 February, I had completed some work in my Nissen hut orderly room and decided to walk across the wire mesh runway to the GCI located in the middle of the drome.

I was almost at the operations room vehicle when I heard the thunder of low flying noisy aircraft most unlike the pleasant singing hum of the Spitfire's Merlin engine. Though the Abbeville boys of the Luftwaffe were frequent early morning visitors to my station at Bolt Head, I was not perturbed as the engine noises came from inland. I turned and, flying straight towards me, were six aircraft of a type that I had never seen before. At a very low altitude they thundered overhead heading south and out to sea, and I recognised them from our aircraft recognition charts to be torpedo carrying Beaufort bombers of the Fleet Air Arm.

They were keeping east of the Plymouth gun area with its trigger-happy naval gunners and by flying low across the sea they seemed to be intent on avoiding pick up by the German Freya radar at Cherbourg.

After a few seconds of mental paralysis, I realised that something special was afoot and I completed the last 100 yards to the operations vehicle in record time.

The crew were still quietly going about their watch duties and were not perturbed by the unusual RAF visitors. On the telephone tie line to our Sector Operations room, I spoke to the controller to find out what the Beaufort aircraft were up to, but without any success. Still under the impression that something important must be happening, I spoke to the CH radar at West Prawle, who had a direct plotting telephone line to Fighter Command headquarters at Stanmore.

It was whilst I was making my enquiry that a big black shipping plaque was placed on their plotting table. The *Scharnhorst, Gneisenau* and *Prinz Eugen* had not only broken out of Brest but had already reached Dieppe. The Beaufort bombers were obviously on their way to make the battleships an unwelcome visit.

The news was stunning. With all the surveillance by our airborne and ground radars, and despite all the months of bombing that had cost us dearly, together with the constant patrolling of the harbour at Brest, which had kept squadrons of aircraft busy for months, we had been caught napping.

The defeat was a body blow to 60 Group, the Radar Group responsible for all British radar, and the voicing of theories by the non-technical staff of our station on what had gone wrong went on well into the night.

On retiring, I usually indulged myself in my luxuriously remote tiny bathroom. I found it an excellent way of escaping from the chatter of the operators who, though skilled at their own job, had no idea what might or might not have gone wrong at Brest.

I knew the detail of the three separate patrol lines as this was our normal operations area. The regular 'Stopper' reconnaissance flights had taken place for months. In fact, I had flown down to St Eval, a little way to our west, to help set up a makeshift ASV beacon for the special use of the aircraft carrying out the surveillance flights. The fact that the beacon could be seen from almost 60 miles out to sea allowed them to easily home on this treacherous southwestern tip of England, enabling the aircraft to spend longer periods on patrol. In the past, aircraft had missed that tenuous extremity of southern England and ended their days in the watery unfeeling depths of the Atlantic.

When working on the ASV beacon in St Eval, I had become quite friendly with the limited technical staff of the station. I felt that I could ask them unofficially for information on what might have gone wrong at Brest, whereas I could not approach anyone officially in Coastal Command.

The brotherhood of technical people was a wonderful network in time of war. I decided to try to make contact with my friends on the morrow.

With Whitehall breathing down their necks, I found it quite difficult to get through to my electronic technician friend in St Eval. Finally, in guarded tones, he informed me that it was the old story. Aircrew when looking for a reason to abort a mission would blame either the aircraft or its equipment.

They had reported three ASV sets as having broken down once in the operational area. This was a hard pill for the technical personnel to swallow as the serviceability record on the excellent and simple ASV radar was impeccable. Nevertheless, all the equipment had been tested in the aircraft on landing, and no fault of any kind was found. One fuse had been removed by an operator and when tested was found to be intact.

One radar operator's report, however, supplied me with the missing clue. He stated that the screen was full of hash showing that the aerial split motor was faulty and that the contacts needed cleaning. There was no sign of this fault when the aircraft was ground tested.

This, however, was the information that made me come up with a startling theory and, realising the import, I kept it strictly to myself.

On instructions from the very top, which meant Churchill himself, all the equipment was being removed from the aircraft for bench tests. This was quite a lot of hard work for the very limited technical staff who felt quite sure it was the 'aircrew' syndrome, and they were going to be the scapegoats. For them it really was Friday the 13th.

I contacted my friends a few days later and the story was the same; the equipment was in good shape. Group had decreed that the sets had shown an intermittent fault, but no faults of any kind were ever found. Because of the furore in Whitehall, some of the equipment was kept on test for weeks, but of course with the same result.

Back at Bolt Head I gave a lot of thought to the whole mess and came to the conclusion that the ASV radars unknowingly had been jammed. The unexpected use of this particular form of jamming had fooled our operators completely.

When I considered what the Germans had achieved by the surprise use of high power, what we would now call 'white noise amplitude modulated continuous wave' jamming, because that is what the Radar Operator had described, I immediately realised that this was what Sergeant Ted Cole had seen on the night of 11 February.

As the German night fighter jamming aircraft was being positioned by a Cherbourg Freya behind the Lockheed Hudson, the radiation at times pointed briefly in our direction.

I became very excited as I realised this might be the way to place ships close inshore during an invasion of the Continent without being detected. It was too simple. Secretly manufacture a lot of high-power jammers for the radar frequencies in question and use them on a once-only basis.

There would have to be complete secrecy, for if the enemy got wind of the plan they would do as we had done in 1938, when we had struggled with the prospect of Mandrel jamming; they would prepare a counter-countermeasure.

I prepared a long screed stating my suspicions and the possibilities. When it was finished, I was not sure whom to send it to. I might well be told by Wing HQ to mind my own business and concentrate on my own work. Placing it in a personal secret file with other material I had prepared, I carried on with the workaday business of running my station.

Naval problems were not my province.

Letter to Tait

The following was the substance of the ultra-secret letter I sent to the Director General of Signals, Air Group Captain Commodore Victor Tait in early March 1942.

> The critical and most difficult part of the operation was to get the ships out of the harbour, negotiate the headland at Ushant and head into the open waters at the mouth of the English Channel.
>
> If that could be achieved a little of that luck which is only the reward of the very courageous would carry them through.
>
> In my opinion they (Martini) had placed a very powerful radar noise type jammer on the tip of the Western Cape. This would have completely jammed the Hudson's ASV radar at this, the most critical point. They might have also prepared and installed a high power 200-megahertz jammer of the same type aboard a German night fighter.
>
> I say this because for short periods my GCI operations crew reported unusual interference on that night. Due to the long ranges involved this could only have been received from an aircraft.
>
> This was the night fighter which had been observed by the Hudson aircraft.
>
> Using the night fighter methods we had both developed for aircraft interception, it was easy for the German night fighter to keep its distance and tail the slow flying Hudson aircraft.
>
> Though we did not know this at that time, the German night fighter's own Lichtenstein AI radar worked on 600 megahertz and the 200-megahertz jamming would not interfere with it.
>
> Of course, the German night fighter could have easily homed in and shot down the Lockheed Hudson, but this would have been self-defeating.
>
> The interference which had never been seen before by the RAF observer would have indicated a faulty aerial split motor on the ASV or a noisy front-end tube in its receiver.
>
> It was certainly enough to cause the RAF aircraft to abandon the 'Stopper' patrol and return to base during that critical period, leaving the ships to head east into the open waters of the English Channel.
>
> Furthermore, it would have been simple to arrange for a shipboard jammer installed on one of the ships to be switched on as it left harbour and pointed in the general direction of the RAF Hudson. This would further ensure the non-operation of the airborne ASV.
>
> A land-based 200-megahertz jammer installed on the Western Cape especially for this operation would have prevented the pickup of the ships at that most critical point when circumventing the coast at Ushant.
>
> Having caused the Hudson aircraft carrying out the 'Stopper' patrol to abort, the ships were virtually home and dry.
>
> 915592 J. Nissenthall, Flight Sergeant.

The jamming had fooled our aircraft radar operators completely. They were convinced that their ASV radar was faulty and the long journeys to and from their base under radio silence had given the German battle cruisers many hours start on their hair-raising trip through the English Channel.

The rest is history.

Despite all our efforts and terrible losses as hasty attacks were improvised on the battleships, the Luftwaffe, by the skilful use of applied countermeasures psychology, had won a major victory.

The lesson was obvious, and it was all too simple: manufacture super high-power jamming equipment for the radio bands you wish to jam, but keep it under wraps.

As we subsequently discovered, under the protective cover of JU88 night fighters arranged by General Galland, the ships steamed at full speed throughout the night covering the 300 miles to Dieppe, where they were finally detected mid-morning on 12 February.

For some time, the excellent high cliff top CHL radar at Beachy Head had been plotting the fighter aircraft escorting the battle cruisers in the mistaken assumption that that they were part of an air sea rescue operation. These tracks were passed to Fighter Command Headquarters at Stanmore, but it was some time before anyone noticed that the group of plots was moving east at the speed of 30 miles an hour.

The CHL, which normally swept like a radio lighthouse, illuminated the ships for only a few seconds of each sweep, which lasted for about a minute.

This could easily be monitored by radio receivers aboard the battleships, and it indicated that the true identity of the plots had not as yet been recognised.

But when, at Stanmore's instruction, they inched the beam slowly backwards and forwards over the ships, the Germans knew that the game was up. They then switched on their high-power jammer which completely wiped out the Beachy Head radar. Fortunately for Wolfgang Martini, both the Beachy Head CHL radar and our airborne ASV radar worked on the same frequency band. The same jammer was effective on both types of equipment.

My suggestion to the Director General of Signals was that by very secretly preparing super high-power jamming equipment, which had never been used before, and then using it on a once-only basis, they had completely defeated our whole superb radar detection organisation.

It was a perfect demonstration of the psychological application of a technique that we could well copy and use for our invasion landings if and when they eventually became possible.

This of course was how the Normandy landings were carried out with very low casualties.

The idea was to prepare a vast quantity of the powerful jamming apparatus and use it on a one night only basis, the night of the invasion. By keeping our original invention and proliferation of radar religiously secret, we had secured an important advantage during the daylight Battle of Britain. This was because the Luftwaffe had not prepared any countermeasures.

History was repeating itself.

Once again, if we could keep the manufacture of the high-power jamming equipment secret, we would be able to give them an unpleasant surprise on the night of the invasion.

The professionals of the Wehrmacht had shown the way.

Possibly, this is what Churchill was referring to in his memoirs when he wrote that in some very mysterious way known only to our most secret circles had the escape of the *Scharnhorst, Gneisenau* and *Prinz Eugen* proved of great advantage to the Allies.

Ten days after the Brest affair, when we were still smarting as a result of the escape of the German battle cruisers, the morale of the people of Britain received a tremendous fillip: the raid on the Wurzburg radar at Bruneval in northern France.

I was thrilled at the accounts of the skilful airborne attack and the withdrawal of the troops who were brought back by Mountbatten's favourite commando carrier, the *Prince Albert*.

For many months I had been receiving and examining the pulses from a radar located on the Cherbourg Peninsula, which operated on the long wavelength of 2.5 metres. It was a Freya radar and for some time I had been intrigued by thoughts of one day paying a visit to a German radar.

There seemed to be no way of this happening, but quite suddenly Combined Operations Command, under the new Chief, Vice Admiral Lord Louis Mountbatten, had shown that nothing was impossible. Immediately I contacted Squadron Leader Charles Davie Keir, the Senior Technical Officer at 78 Wing Headquarters in Ashburton and volunteered to take part in any future raids on German radar stations. Furthermore, I suggested that the Freya located at the tip of the Cotentin Peninsula might be a worthy target.

Squadron Leader Keir suggested that I write a personal letter to the Director General of Signals in Whitehall, offering my services.

He said that the letter could be sealed and that nobody would be privy to its contents other than the Director General of Signals and myself.

This was a wonderful opportunity. I prepared the letter volunteering my services and in it I included my letter on what I felt had actually taken place on the escape of the German battleships. I then made the point that this might be the way of approaching the coast of France without being blown to smithereens.

A week or two later, I was surprised to receive an attachment to an Air Ministry Unit at Long Cross. This turned out to be a tented camp and acted as a 'safe house' located near London and just outside Egham.

From Long Cross I received transport which took me to the Air Ministry, King Charles Street, Whitehall. I then discovered that I had been booked for an interview with Air Commodore Victor Hubert Tait (later Air Vice Marshal Sir Victor Tait), with whom I was to form a lifelong friendship. Tait was a Winnipeg Canadian who was the RAF Director General of Signals, the man who masterminded British radar throughout the Second World War.

I had never been summoned to the Air Ministry in Whitehall and was awed at the opulence of this Victorian building, with its huge windows, high ceilinged rooms and elegant desks.

I reached the huge oak door marked Director General of Signals and, after making sure that my tie was straight and my forage cap in its correct position with the two small buttons 1 inch over my right eye, I knocked on the door.

A pleasant voice bade me enter. I shut the door and marched to the front of the large oak desk and saluted RAF style, long way up and short way down. I gave my name and number and waited at attention.

For quite a time the tanned Air Commodore stared at me unsmiling and without speaking. He was, I feel, trying to sum me up and, apparently satisfied with what he saw, his face broke into a broad grin, and he told me to remove my hat.

Leaving his massive chair, he came round to the front of the desk and shook my hand. I was surprised and relieved as I did not expect this informal treatment from such a senior officer. Despite his long sojourn in England, he was still a typical pleasant Canadian.

We sat in two huge luxurious leather armchairs and very quickly he made me relax and feel at home. After asking some personal questions about my widowed mother, he quickly got down to business.

I could see that the various suggestions I had made over the last two years and the drawings that accompanied them were neatly attached to my Form 1580 in a file on his desk. This form is really a report card on your activities in the air force.

He told me how he had sent Flight Lieutenant Hunt up to my unit in Rosehearty in Scotland early in 1940 to discuss the suggestion I had made about a VHF beamed communications system. I had stumbled on a project which was already under consideration for development. Hunt had asked me not to discuss any of the detail with other technical people on the station, as secrecy would be paramount when it finally came into use.

For technical buffs, I had merely suggested that an Armstrong frequency modulation signal after the limiter was merely a train of square waves of different sizes. My suggestion was to develop a modulator to turn speech into trains of these square waves and transmit them on a VHF beam exactly like our CHL.

This, of course, was part of the basis of the No 11 set, which much later in the war was used for long distance secret communications by General Montgomery. This is why he always made his field headquarters on high ground as he made his way across France and Germany.

Tait also mentioned one or two other ideas in my file. I was amazed at how familiar he was with my propositions. It should be remembered that I was a lowly Flight Sergeant and he an exalted Air Commodore Group Captain. Eventually we got to my recent letter that included the offer to take part in a raid on occupied France to investigate the Freya radar.

He told me that my offer would be considered, but not my choice of target located near the tip of the Contentin Peninsula, which I had been monitoring for more than a year.

He then became far more serious when discussing my theories on the escape of the German battle cruisers *Scharnhorst* and *Gneisenau*. Apparently, he had come to the same conclusions on the escape and agreed with my suggestion that we could apply the same methods to get our ships close inshore on the day of the invasion when and if it ever arrived.

I had made the point that the plan was ridiculously simple, but it would only work if infinite secrecy on our intentions could be maintained. The jamming equipment would have to be designed and manufactured with all those involved thinking that it was to have some other application, for example bomber protection. If there was idle and excited chatter amongst our radar fraternity and the civilian scientists, it would get to the ears of the Germans, and they would simply prepare counter-countermeasures.

He instructed me not to discuss my ideas with anybody else and that I could contact him at any time. I was thrilled when he told me I was now in a select circle which included the Chief of Combined Operations, who as we know was also a radio specialist.

On subsequent visits to Whitehall, I was allowed to spend time in Richmond Terrace, the headquarters of Combined Operations, where my ideas and suggestions were gratefully received.

My mother, who had been bombed out of our house in the East End of London, appreciated these odd trips, as I could always spend the night with her and my younger brother at our new home in Ealing, West London.

Victor Tait introduced me to Mountbatten and in fact the CCO (Chief Commanding Officer) was to personally brief both me and Chief Petty Officer Kenneth Dearson, the second-best telegraphist in the navy (Mountbatten regarded himself as No 1), for the electronic intelligence assault on a German Combined Operations Plotting and Reporting Centre, at the lighthouse Phare D'Ailly, some 5 miles west of the French port of Dieppe. As we will see, the resulting communications bonanza carried us through to D-Day and beyond. It was to be the most secret and wonderful outcome of the raid, carried out later that year in the general area of Dieppe.

Air Commodore Tait underlined the need for absolute secrecy on anything that we discussed or anything he proposed to me.

Not that the need for secrecy was anything new, as all my radar activities since before the outbreak of war were cloaked in absolute secrecy. Even my own family had no conception of anything, other than I was involved in some wireless work for the RAF.

The very letters RDF were a sacred trust and never uttered in public and as a result, the Luftwaffe, when it reached Calais in June 1940, discovered to its horror that we had a complete integrated system of air defence.

The Seetakt, as inspected on the scuttled German battleship the *Graf Spee* early in the war, was a short-range gunnery radar. The Wurzburg as a result of Operation Biting, the Bruneval Raid, which had taken place at the end of February 1942, was also a short-range gunnery radar

Now Tait wanted a firm evaluation of the third German radar, the Freya, which he wanted to take place fairly quickly.

CHAPTER 16
DIEPPE

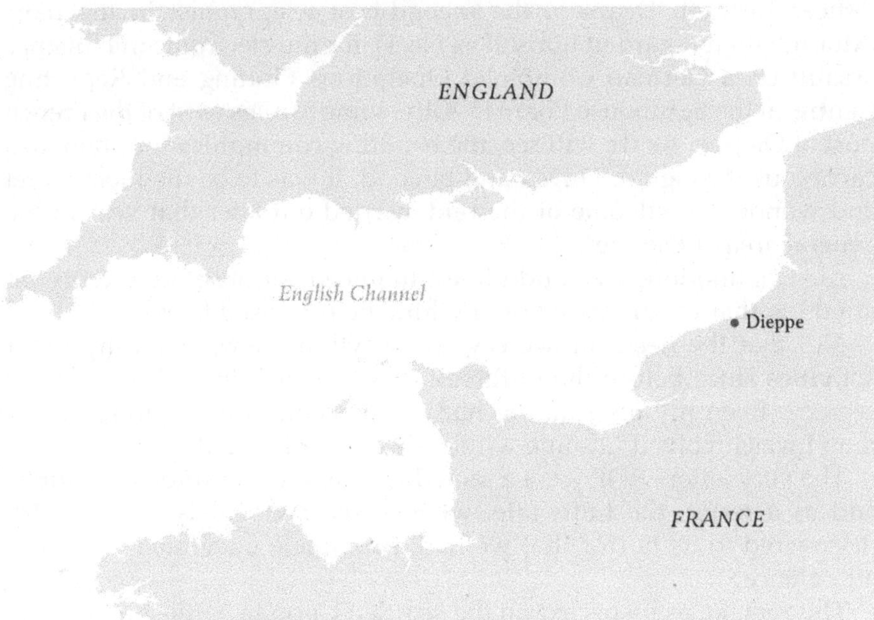

Dieppe Raid, 19 August 1942

In mid-March our group agreed that one of the targets of the next raid would be the 'destruction and evaluation' of the Freya radar near Dieppe. Secretly, my object, of course, was a verification of the fact that it had now become a precision radar, if that were so.

With the complete success of Operation Biting as a guide, it was suggested we merely repeat the medicine at Pourville.

There was one major snag. Apart from the local defences, we would be landing less than 3 miles from the huge defensive garrison at Dieppe. In quick time, they would be in Pourville.

A larger force was suggested, but this would have led to a difficult and dangerous evacuation under fire from the guns at Berneval-le-Grand and Puys, east of Dieppe.

The low-level discussion group then suggested two large landings: a large force including airborne troops to neutralise the guns to the east, and another force to land at Pourville.

To keep the Dieppe garrison at home, a 'feint' convoy would approach the main Dieppe beach front, and this would be clearly viewed by the Freya to be visited. The feint convoy would then withdraw once the evacuations started from the flanking beaches. With all these ships milling around during the evacuation of the troops, the guns of another battery 3 miles to the west, at Varengeville, could also ruin the evacuation. These would also have to be taken out by an airborne force. The expanded plan became singularly exciting and attractive as apart from the Freya at Pourville, there was a Wurzburg radar near Puys and a Seetakt radar and a whole combined services plotting centre in the Phare D'Ailly lighthouse complex at Varengeville.

As there was a strong beach defence position near the lighthouse, a squadron of three tanks was mooted as there might be heavy resistance to the landing.

That seemed to be the situation when I had to break off and return to my station: three Operation Bitings on the flanks and a feint convoy to keep the German garrison at home in Dieppe. It all sounded most exciting.

Mountbatten always kept Churchill appraised of the planning discussions. The Prime Minister decided that this was no longer a raid as tanks were to be involved. This added another dimension and Churchill suggested that an army advisor would have to help in the planning. As a result, the Chief of the Imperial General Staff, Sir Alan Brooke detailed the GOC of Southern Command General Bernard Montgomery to assist and advise the Combined Operations Planners.

In early June I was called to Whitehall by the Director General of Signals. On arrival I discovered that it was a primary briefing on a proposed operation, codenamed Operation Rutter, on an area to the west of Dieppe. Chief Petty Officer Kenneth Dearson, who was stationed on the commando carrier communications ship *Prince Albert*, was called to Whitehall for the same briefing.

Mountbatten instructed us to work together as a team, Dearson, as his top radio man and I, Tait's man, representing the RAF. The meeting was held under the most stringent security, and the only people present were the Chief of Combined Operations, Admiral Lord Louis Mountbatten, Air Commodore Victor Tait, Captain Ward of the *Princess Beatrix,* Captain John Hughes-Hallett of Combined Operations and Wing Commander the Marquis De Casa Maury, the Senior Intelligence Officer of Combined Operations.

Dearson and I were read certain provisions of the Defence of the Realm Act by the Senior Intelligence Officer, whose cultured English did not seem to accord with his Spanish title. Generally, however, the whole meeting was conducted in a very friendly atmosphere considering the wide divergence in rank of the participants.

The only person who seemed to object to the presence of two non-commissioned officers was the SIO, Casa Maury. After it had been pointed out by the Chief of Combined Operations that we, the two Non-Commissioned Officers, were the cutting edge of the operation, he had to contain himself and his sarcasm.

The fairly large military operation was to be carried out in an area where there were two German radars, a Freya at Pourville, near Dieppe and a Seetakt at the German Combined Plotting Centre at the lighthouse at Phare D'Ailly, about 3 miles west. The whole of the coast of France it seemed was a series of German *Stützpunkte,* standard self-contained defensive fortresses. Hence 'Festung Europa', Fortress Europe.

Our portion of the operation was to carry out a radar and telecommunications reconnaissance to verify how the Germans would go about defending themselves when violently attacked. The object was to learn how to defeat the German emergency communications systems when we eventually launched our full invasion of Europe. It was to be the very first Electronic Intelligence gathering mission.

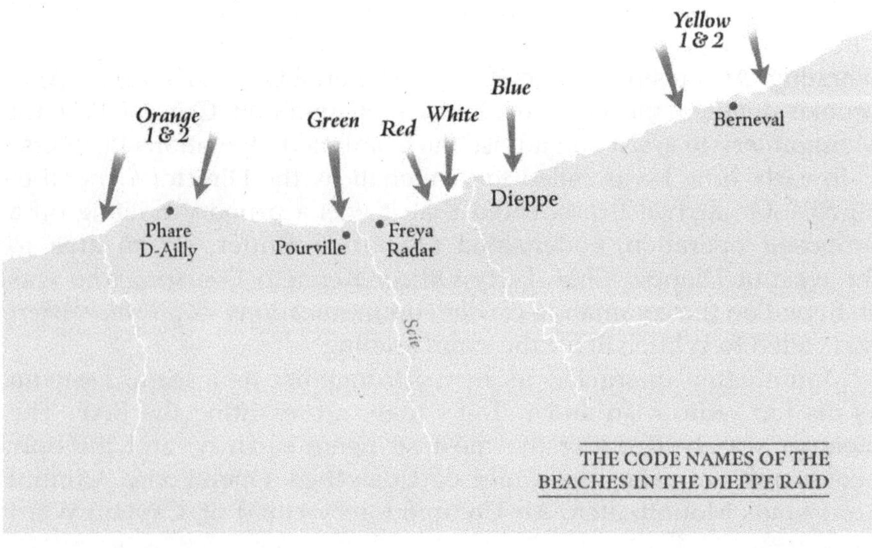

THE CODE NAMES OF THE BEACHES IN THE DIEPPE RAID

Our part of the proposed operation would take place over two beaches, Green and Orange. I would go ashore with the special service troops. At this time these were Commandos, and the object was to verify a certain important detail with respect to the Freya. I would have a team who would have to help and protect me and make sure that I was not captured alive. Whilst ashore, we as a team would have to damage line communications everywhere in an effort to make the Germans revert to their standby emergency radio communications if any.

Dearson would be aboard the *Prince Albert* with a selected team of German speaking wireless telegraphy operators, some very special radio receivers and a new sound recording machine 'borrowed' from the USS *Augusta* for recording radio telephone transmissions. With this equipment Dearson would be able to verify the transmission modes and the radio frequencies used by the German army, navy and air force. After staying the night with my mother at her home in Ealing, I returned to Bolt Head and my normal night fighting duties.

For Operation Rutter, on 4 July 1942 I left my night fighter control station, RAF Hope Cove near Plymouth, and made my way to Whitehall where I picked up my specialised equipment. I was then escorted by an SOE Lieutenant to Havant where we boarded the ferry to Ryde.

After being handed over to the Commanding Officer of the South Saskatchewan Regiment, I took up my very temporary residence at Osborne House. This magnificent building had been the summer residence of Queen Victoria and, of course, Mountbatten had spent much of his youth in the beautiful gardens of this very elegant mansion.

The next day together with the men of the South Saskatchewan Regiment, I boarded our infantry landing ship, the *Princess Beatrix*, which was anchored out in the Southampton Water just off the Isle of Wight. This was at the beginning of July.

We lay in a very exposed position quite near the coast of France and I was on tenterhooks as the Luftwaffe was a frequent visitor to this part of the world. Captain Lang, also worried, asked me to get the ship's radar working so that he could run his own defensive radar watch.

On the following morning, Mountbatten travelled from ship to ship to wish the troops bon voyage. Finally, he was piped aboard the *Princess Beatrix*. With his entourage he climbed up to the bridge where we were all lined up for his visit. In his naval uniform with gold braid right up to his elbow, he looked magnificent and exuded confidence and professionalism. Captain Lang introduced us one at a time and told Mountbatten how he had already put me to work on the ship's radar.

Mountbatten found this amusing, and we spoke briefly on the worrisome radar topics as they concerned the raid. After all, I was going to visit the radar that was almost definitely going to plot us for some time and certainly long before we made landfall. He pointed out that in the fog of war, nothing worked out the way it would seem.

He predicted precisely what might happen in the worst instance and in fact this is what exactly took place. He then shook hands, wished me good luck and finished by putting both his hands on my shoulders and stating simply, 'You will be alright'.

How right he was.

He then took the microphone and, using the ship's loudspeaker system, spoke to the men assembled on the deck.

The men raised a cheer, and he then made his way down to the launch. He was piped off the ship and made his way to the next vessel.

Mountbatten was a born leader who inspired supreme confidence, and after his visit the men were ready for anything.

On the other side of the English Channel the weather was bad, and the operation was delayed. On the evening of the second day, a German reconnaissance aircraft flew over the assembled fleet and promptly at 6.15 next morning, we were bombed. There were no casualties, but the *Queen Charlotte* and another ship were damaged. Operation Rutter was cancelled. The disappointed Canadians were returned to their camps in Sussex, and I returned to my radar station near Plymouth.

The night of 17 August had been the usual tiring round of practice interceptions over the Channel, and I had been quite glad when, in the absence of enemy aircraft activity, we were told to stand down.

I was looking forward to a good morning's sleep back at the hotel when my orderly room sergeant Bill Powell, who had just come on duty, phoned my operation room tender and reported in his amusing sarcastic style that 'My limousine was awaiting without'.

Once again, under the cover of an Air Ministry Attachment signal, I had to leave for London. This time we went straight to Whitehall, where Victor Tait told me that the operation had been resuscitated.

We stopped briefly at Richmond Terrace, not because I enjoyed visiting Casa Maury, but to pick up my special equipment, tools and escape set which were kept in Casa Maury's office in a special lockup.

With a Special Operations Executive escort, I was then driven down to Crondall, where a Marine headquarters building was located near the village. There I was given the sad news by a Marine Major that the plan had been changed and that the armour support I expected would not materialise. It sounded like a death warrant and if we met any resistance, we might never get to the top of the cliff where the radar was located.

Now, with the Marine officer on board, we made our way to Southampton.

Our papers were checked by a sailor as we entered King George V Dock and we drove along the dockside to the gangplank where I boarded the ship, which had been waiting for me. The gangplank was snatched away, and we cast off immediately.

It was a beautiful summer's evening and at high speed the crew rigged up a dummy funnel and hoisted the neutral Spanish flag as we swept out to sea. It was still light enough for a German reconnaissance to spot and report our movements and the disguise would hopefully help to fool any German snoopers.

We swept along the coast where I had a beautiful view of the sunny sandy beaches where I had spent many enjoyable holidays in happier times. We then turned south and took up our position as the lead ship.

It grew dark as we proceeded and leaning on the handrail, I was very conscious of the flapping canvas over my head. I have always been thrilled at the sight and sound of flags cavorting in the breeze, and I felt very proud and excited at the great adventure that lay ahead.

I was deep in thought when I received a tap on the shoulder. It was Staff Sergeant Roy Hawkins of the Canadian Field Security Service. He told me that he was to join my party and had been instructed to look after me. I asked him wryly whether he was going to show the Germans his stripes. I had no illusions; I realised we were heading into a battle royal.

The captain ordered us to inflate our canvas Mae Wests as we were entering the minefield. We passed the faint marker light on the Naval Type 78 marshalling radar transponder, which marked the entry to the mine-swept zone, and we eventually reached our drop zone some 10 miles off the coast of France at about 2.30 am. The landing craft were lowered down into the sea and rope nets were thrown over the sides of the ship to enable us to scramble down and climb aboard.

The crew started up the two V8 engines and together with all the other craft we were on the way to our target, Green Beach. It was to be slightly less than a two-hour trip and we travelled in complete darkness as there was no moon.

When we had been under way for almost an hour, we were suddenly shocked and blinded by a huge flare that had been fired high in the sky some 5 miles to our east.

I climbed up to look over the armoured side of the landing craft and was horrified to see in the light of the flare a long line of landing craft that seemed to stretch into infinity.

It seemed that as we had been exposed, there could be no thought of surprise.

Instantly heavy automatic gun fire broke out near the flare, and I could see tracer fire going both ways as the ships, whoever they were, fought out their battle. After a minute or two the firing ceased, and we carried on our way to Green Beach.

There were control telegraph bells as we slowed down, and the landing craft turned to starboard and started the run into the beach. The ramp crashed down and we charged out of the craft, staggering and fighting our way up the steep beach through the heavy shingle to the sea wall.

I looked up at the night sky and slowly made out the outline of some large houses right overhead on the beach front. With shock I realised that the navy had dropped us at the wrong place. We were supposed to be placed further to the east, astride the tiny River Scie which flowed out to the sea on the eastern side of Pourville. Instead, as it transpired, we had been dropped in front of houses billeting the German garrison troops.

The noise we made had awakened them and the fighting started immediately. The unfortunate Germans had been surprised. They fell out of bed and were fighting in their underclothes. Despite everything, we had gained surprise for the landing, but our first traumatic mistake was a serious one.

With no navigational aids the navy had done their best, but we had been landed in the wrong place. I knew how this could be avoided and this was a mistake which must not happen in the future, when a small army might be transported to the wrong landing point.

Apart from some deceptively pretty, orange spheres of tracer that floated overhead, marking the path of small calibre shell fire from the general direction of the radar station, together with the muted rifle fire and grenade explosions which took place when the local Wehrmacht garrison sleeping in houses on the beachfront were receiving treatment, the landing itself was quite peaceful.

Pourville is a small seaside village with a tiny square almost on the sea front. Whilst my group located each other in the predawn darkness, I placed some hand grenades in a telephone distribution box on the corner of the square leading southwest out of Pourville. This I knew was the Varengeville road, which would wind its way towards the lighthouse Phare D'Ailly. This I hoped would help to isolate the lighthouse and disrupt communications, hopefully making them revert to radio communications.

On a tree over the same distribution box were some army 'Don 8'-type telephone wires leading cross-country up towards the radar station. With my cutters, I severed these wires, which I was sure would be used by spotters for directing the fire of guns perhaps further up the hill. Certainly, and thankfully, the subsequent shell fire on the village was more than haphazard.

As Mountbatten had hammered home during my briefing with Chief Petty Officer Dearson, communications were the lifeblood of defence, and accordingly, I had instructed my team to destroy any telephone landlines, military or civilian, that we came across during our short stay in occupied France.

A huge twin concrete roadblock barred our road out of the village and the German gunners manning the guns in the area caused some casualties amongst the men of 'A' Company. We then advanced towards the small bridge over the tiny River Scie. Here there was some problem with the junior commanders and their men who had taken cover on the western side of the bridge.

Finally, Captain Murray Osten, the Company Commander, took command himself and led my small group forward. He found a high-hedged winding lane that led up the hill in comparative safety.

With my own small party of men of the South Saskatchewan Regiment and a few of the others from 'A' Company, we made our way up the long lane leading to the top of the hill towards the radar station.

High up the hill at the end of the lane we reached a T-junction, which seemed to mark the western boundary of the plateau on which the Freya was located. In front of us was a thick hedge lined with barbed wire.

From this vantage point silhouetted against the skyline to the northeast I could clearly see the huge 40-foot gantry rotating as it tracked aircraft movements. It was not sweeping continuously as a CHL but was obviously following tracks of individual aircraft, very much like a modern dish. I did not know how they accomplished this kind of precision on such a large structure, but I had my answer: it really was a precision radar.

A Forward Observation Officer (FOO) with a portable radio transceiver was supposed to accompany my group, but he and the rest of 'A' Company were most conspicuous by their absence.

The plan had been to signal PO Dearson aboard the *Prince Albert* with the simple answer on my findings, but with no radio contact I had to resort to the more dangerous task, which I had discussed with Dr Don Preist of the Telecommunications Establishment at Malvern.

Preist had been the Air Ministry representative aboard the *Prince Albert* during Operation Biting, the raid on Bruneval. At the time of writing (1993) Don Preist is alive and well and lives in Colorado Springs in the United States.

It was his suggestion to cut the telephone wires leading into the station, hopefully making the German plotters revert to standby radio communication.

Dearson was waiting for either my signal from the missing Forward Observation Officer or signals directly from the Freya station.

On the far western flank at Varengeville, a seaborne force of No 4 Commando had replaced the parachute force that was going to spike the guns of the battery situated almost a mile inland. This had to somehow be accomplished without alerting the combined services plotting centre based in the lighthouse at Phare D'Ailly, as this was the real target as outlined in plans first discussed in March of 1942.

What would have been a straightforward task for airborne troops now required a full military pincer movement which had to be most carefully planned and timed as it would have to be launched from the two Orange Beaches with all the attendant possibilities of detection.

Together with the communications intelligence team, the Commandos who had travelled over in the *Prince Albert* landed on the two different beaches.

The plan of attack was devised and carried out by the same man, Lord Lovat.

One troop landed and dealt with the heavily defended 'resistance nest' beach gun positions. They also cut the telephone wires from the nearby Phare D'Ailly lighthouse housing the eighty German navy, army and air force plotting and communications personnel. This disrupted communications to their various formation headquarters in Dieppe and beyond!

Reverting to radio communication for Dearson's benefit, they carried on blissfully unaware of what was taking place around them. In fact, it was almost 10 am before they discovered that the gun battery located a little way to the south had been put to the bayonet and destroyed.

In the intervening period, the *Flugmeldcentrum* (Flight Reporting Centre) in Dieppe had on a few occasions requested information on the general situation in the Varengeville region and the reply was always peaceful and reassuring.

The telephone wires were cut by Trooper Finney, a Commando signals specialist, and the army, navy and Luftwaffe plotters, together with the communications personnel in the lighthouse, reverted to radio communication and Chief Petty Officer Dearson with his team aboard the *Prince Albert* went into action.

They recorded all the radio frequencies, and the modes of communication used for contact with German Army Divisional Headquarters, the Luftwaffe *Flugmeldcentrum* and the naval headquarters in the Dieppe area.

Quietly skirting the lighthouse, one group of No 4 Commando approached for a frontal assault on the gun battery, whilst the other did a wide circle round to approach the gun battery from the rear.

At 6.29 am, when the rear group were in position the plan was to fire a Verey signal high in the sky. The first group would then attack and with the Germans fully engaged, the rear attack would take place.

Engineered by Lord Lovat, it was a classic pincer movement. As ever, Murphy's law prevailed, but in true Commando style, Lovat's men took casualties and persevered with the attacks putting all the gun crew to the bayonet. They then spiked the guns and with some prisoners, withdrew in good order. Though seriously wounded, Major Pat Porteous, whom the author is proud to call his friend, was to earn the Victoria Cross.

Two Spitfires of 129 Squadron opened the proceedings by launching a cannon attack on the large plate glass window of the lighthouse at precisely 6.29 am. This, it was anticipated, would keep the lighthouse inmates occupied just at that time when the attack on the gun battery was going in.

The two aircraft, one flown by Flying Officer Jones and the other by Sergeant Dick Reeves, both of Toronto, made landfall a mile or two to the west and a trifle early for the assignment.

To keep the vital 6.29 am appointment, the two aircraft had to fly fairly low and slow along the coast, the aircraft brilliantly illuminated by early morning sunlight. Sadly, light anti-aircraft fire shot one of the aircraft down, but Sergeant Reeves (later Flight Lieutenant Reeves DFC), kept the appointment at the lighthouse and completely distracted the communications personnel at that precise moment when Lord Lovat and his men went into the attack.

After collecting a veritable bonanza of information on the radio frequencies and modes of operation of a German *Stützpunkte* when under attack, the *Prince Albert,* with CPO Kenneth Dearson and his electronic intelligence team aboard, moved to the east to keep his electronic rendezvous with the author on Green Beach. Escorted by the destroyers *Albrighton* and *Fernie,* the *Prince Albert* then moved 1 mile to the east and lay 2 miles off the River Scie, due north of the Freya radar.

With Lord Lovat now safely aboard, the *Prince Albert,* its mission completed on Orange Beach, moved to the agreed point some distance north of the mouth of the River Scie. From this position radio signals emanating from the region at the Freya would now be strongly received.

Using the very first model of the Webster-Chicago Wire Sound Recorder (a forerunner of the tape recorder) and an RCA AR-88 specially prepared scanning receiver 'borrowed' by Mountbatten from the American cruiser USS *Augusta,* Dearson with his team recorded plotting information and other communications intelligence right up until they were back in port many hours later. With his team of German-speaking wireless telegraphy operators, they also recorded the discomfiture of the Germans as their landline's communications failed for various reasons during the fighting.

By the end of the day, we had learned the operational radio frequencies of all the various German formations, as well as their modes of transmission. This was most comforting as the plaintive unanswered calls for help on weak and unstable transmitters indicated that our adversaries were no way near as erudite as we had imagined. No doubt in time they would improve their network, but on their showing that day we realised that they could be easily jammed by simple equipment.

It was at least a starting point, and this was to prove of great advantage when planning future operations up to and including D-Day. Sadly, at the time of Dieppe, reconnaissance for the invasion of the Continent was little more than a pipe dream.

But back to the hedgerow bordering the Freya station.

Despite Captain Osten's vituperative demand for more support that all but burnt out the tubes in his No 18 set, Colonel Merritt could not get the bulk of the men of 'A' Company to move out of the houses in Pourville and up the hill. For this reason, the few very lightly manned gun posts, including the one firing down on to the west side of the Dieppe beach, were not taken.

The fact that the high ground around the radar station was not cleared had calamitous results for some Canadians who had landed on the western end of the main Dieppe beach.

Thoroughly frightened, I crawled crocodile style along the road at the rear of the station, becoming braver as I proceeded. I had not been observed simply because there was nobody around and cutting the telephone wires going into the station was not difficult.

There were at least two heavy machine gun posts on the eastern side of the Scie valley, but these were emplacements with all fire directed west towards Pourville. One was downhill and near the cliff edge and could enfilade the Pourville beach. Another on a grassy slope slightly inland was an underground emplacement that could only fire in a westerly direction. Firing on fixed lines this made the bridge crossing of the River Scie somewhat hazardous.

From my vantage point on the telephone pole, I had an amazing view of the western side of the operation.

To the east over the Dieppe beaches there was a huge pall of smoke and the thunder and crump of bombs and shells. I had no idea of the drama being played out down below. Looking west, I had a beautiful vista of the Scie valley, with fighter aircraft way below engaged in combat with aircraft flying low and slow, trying to position themselves behind each other in order to deliver a lethal squirt.

High, and not so high, overhead aircraft were in combat with the staccato racket of aircraft cannon and guns and every so often a parachute would blossom out with its human appendage dangling and swinging below.

In a few hectic hours of fighting over the five beachheads, the RAF were to lose more than 100 aircraft and the Luftwaffe about half that number. RAF losses were far greater than on any day of the 1940 Battle of Britain.

The German defences had all been set up to deal with a landing at Pourville village and there were no German troops on the high ground or behind the radar station. The road leading east to Dieppe town was also deserted. Obviously, the defence force had not expected anyone to succeed in overcoming the large village garrison and advancing all the way up the hill. Whilst I was busy up the telephone pole, I was shocked when an automatic weapon opened up nearby. I am sure, however, that I was not the target.

After seven hours ashore I managed to get down to the beach and swam out under heavy small arms fire to an immobilised landing craft, which lay a considerable distance offshore. It was being protected from two predatory fighter aircraft by smoke being laid by a destroyer which had stationed itself slightly further out to sea. The only other survivor of this hair-raising escape was Staff Sergeant Roy Hawkins of Fort McMurray, Alberta.

When the ALC became serviceable, we made to the destroyer *Albrighton* that eventually brought us back to England.

After a rough interrogation by debriefing officers of the Canadian army, which lasted into the early hours of the morning, Roy Hawkins, himself a Field Security Sergeant, managed to secure my release without my breaking cover and explaining my presence on the operation. Having been detailed to accompany me at the very last moment, he had no idea what it was all about. He had received his orders whatever they were, and part of them were to keep me company on this expedition.

It was pretty obvious to him that I was no spy as my interrogating officers were suggesting. After all, we had landed in France just after 4.30 in the morning and had spent seven hours touring the French countryside and being shot at most of the time.

Eventually, after midday, we had swum out under heavy fire from the German troops, which surely suggested that I was no admirer of Hitler.

It was now about 2 am and I was utterly exhausted, not having slept for almost two days. My face was red raw and sore from saltwater spray, privation, mortar blast etc., and I had a huge painful gash on my cheek which I had received when a machine gun had eventually opened up on Murray Osten and myself at the radar station.

Despite my unspeakable condition, the Canadian debriefing officers gave me a merciless interrogation with all kinds of threats in an effort to extract a 'confession'. As the acronym RADAR had not as yet been coined, and the letters RDF were sacrosanct, my apparent obstinacy in refusing to give any reason for my inclusion in the operation annoyed the interrogating officers.

Finally, with Roy Hawkins' help, I signed a note stating that we had tried to gain entry into a German radio station but had been repulsed. (This note I understand is now in an Ottawa Museum).

With my battle dress trousers in ribbons, wearing a borrowed Royal Marine battledress tunic and carrying a helmet which was out of shape and would not go on my head, I was finally released by my inquisitors.

Armed with a railway warrant and taking an early morning Southern Railway's workman's train, much to my fellow passengers' disdain, I made my way with great difficulty to the Air Ministry in Whitehall.

Despite the early hour, Air Vice Marshal Sir Victor Tait was already in his office. Naturally he was overjoyed to see me as I had been reported captured at 12 am on the previous day. At 2 pm this had been replaced by another message from Wing Commander Casa Mauray, the Combined Operations Senior Intelligence Officer, stating that I had been killed in action.

I was not sorry to disappoint him.

I could now tell the Director General to his face that he was correct: the Pourville Freya was indeed a precision radar. Some time later, we discovered that the early Freya radars used the maximum response method to obtain the bearings on an aircraft. This gave an accuracy of about five degrees.

Starting from June 1942, fitting parties under the orders of Wachtmeister Christiaan Ganzer were modifying the Freya radars in northern France to give them a greater degree of bearing precision. Using the split beam method of direction finding, they achieved a bearing accuracy of better than one degree, earning their new title of Freya A-N. Ganzer informed me that Freya 28 at Dieppe had been one of the first to be modified, turning it into a precision radar. Of course, I only received this information after the war.

This vital positive intelligence received as a result of the raid indicated that money and expensive research time could now be confidently invested in the design and manufacture of the appropriate jamming equipment. It would be centred in the first place on 125 megahertz. A low-power version of this equipment, codenamed 'Mandrel', went into service for Bomber Command about two months after the Dieppe Raid.

It was this device more than any other that made the invasion of occupied France and the liberation of the enslaved peoples of Europe a feasible proposition.

After Operation Jubilee, which was a costly disaster in terms of casualties and loss of materiel, there was a lot of soul searching.

There were hectic discussions on the shortcomings of this, the first trial invasion of the continent of Europe, and it became apparent that many of the participants now had positive and outspoken constructive criticisms of every aspect of the operation. The successes, such as the landings on Orange and Green Beaches and the Electronic Intelligence fall-out for various reasons had to remain secret.

As with all battles deemed lost, there was a welter of argument, but Mountbatten had a small army of experienced blooded Combined Operations personnel who were now already contributing so much new 'know-how' to the ongoing and future operations of the ever-expanding Combined Operations Command.

The American generals stationed in London were singularly quiet, having now for the first time had a clear demonstration of the fact that contrary to their opinions, landing in occupied France was going to be a hazardous venture.

Prior to the Dieppe Raid, many subjects that had been the topics of much discussion and varied opinion by the series of committees at Combined Operations Headquarters, were now stark realities.

The primary lesson of the Dieppe Raid was the need for a shattering, drenching artillery bombardment to pulverise and terrify the defending troops. There could be no argument about this vital requirement for any landing in face of a soundly established defence force. This would have to be carried out by a powerful bomber force and a naval bombardment to blow up the local gun positions leaving the defending troops stupefied and stunned. Also, beachhead artillery support was needed for the men carrying out the initial assault. There would have to be far more LCGs, beach landing craft fitted out with guns of at least 4-inch calibre to instantly deal with local resistance nests.

The destroyers had performed magnificently in this capacity, but to avoid running aground they had to keep their distance when attempting to assist the hard-pressed soldiers.

The artillery would have to be on the beach in the initial landings, where at short range they could take up the battle in support of the hard-pressed sappers.

Many more beach landing gun craft were urgently needed. On the main beachfront at Dieppe the devastating fire of the cannon firing fighters had initially been very effective and the German gunners had gone to ground when attacked, but in ten minutes, once the smoke had cleared, they quickly recovered and, as the supporting tanks had not arrived on time, the brave sappers and engineers were suddenly exposed and left stranded in the barbed wire where they were literally torn to pieces.

As mentioned, what was really needed was the bombardment power of an attack force of battleships. In the shallow waters and narrow confines of the English Channel this was obviously an almost impossible requirement. After the Dieppe Raid and in an attempt to answer this question, the cleared deck of a Tank Landing Craft was experimentally fitted with a battery of just under 1,000 rocket launchers. The TLC had to point its 'bow' in the direction of the beach to be 'drenched with fire' and when at the appropriate range, its rockets would be ignited. As each rocket packed the power of a 6-inch shell, it was hoped that it would give the impression of a large naval attack force. This would have the requisite resultant demoralising effect on the defending troops. The rub was that the ranging would have to be accurate or the troops on their way to shore might be decimated, as would any other small craft that accidentally got in the way. In the poor predawn light, how could this possibly be achieved?

As with the ASV Mark II that had been belatedly fitted to destroyers earlier in the war, the RAF once again came to the rescue.

In Scotland a test was carried out using the workhorse ASV II, but its four-microsecond pulse width was far too large.

This meant that due to the short range of the rockets, which was less than a mile, the fall of the shot was masked in the transmitted pulse and the local radar ground returns. Fortunately, at this time the microwave radar H2S was practically ready for use by Bomber Command, and this could have a pulse width of one microsecond or less.

Fortuitously, the equipment which could work off the landing craft's power supply was completely suited to the job. There was plenty of room to stow the twelve boxes holding the apparatus and the results were immediately phenomenal.

When mounted high on the stern, the H2S scanner gave an all-round picture of all the ships in the immediate vicinity, including even the smallest motor launch and marker buoys. The whole flotilla of Landing Craft Assaults speeding towards the shore could now clearly be seen as well as the shoreline of the beach. This was vital as rockets are a fixed range weapon. The Landing Craft Rocket, LCT(R) as it was called, had to steer directly at the beach to be assaulted and when the H2S PPI screen showed a range of 1,700 yards to the target, the rockets had to be ignited.

To Mountbatten's delight the RAF radar turned the highly dangerous rocket-laden craft into a feasible and successful weapon of war. Once again it was the Dieppe experience that brought this fire-breathing monster into being, as a way had to be found to drench defending troops with a soul-shattering blast before the troops landed.

The H2S also displayed all the high points such as cliffs and buildings with great clarity. Another plus was that being a microwave radar, it was not in any way affected by our high-power Mandrel jamming and could not be detected by German coast-watching electronics intelligence snoopers. It was the perfect rocket-sighting device for this new and frightful weapon of war, and it was to be used with huge success in all Combined Operations beach landings.

Another problem brought to light at Dieppe and that needed urgent attention was an analysis of any beach on which landing was contemplated. The Combined Operations training required certain conditions to be existent if beach landing craft were to be able to beach, drop their troops and quickly kedge themselves off the beach. First and foremost, it had to be a real beach.

So many apparent beaches have a large sandy shelf some yards offshore. Landing on these would result in the 'drowning' of vehicles and personnel. These false beaches had to be discovered in good time and avoided as the wastage of precious landing craft would have a deleterious effect on the continuity of the subsequent operation. A shelved beach could make a landing craft broach to, blocking the beach, making withdrawal very difficult, but even more to the point, blocking the passage of other landing craft.

All Combined Operations planning and training required a quick turn around after a beach landing craft had dropped its load. This was the only way in which a large number of troops and materiel could be landed quickly enough to deal with an enemy reaction. There were all kinds of beaches, some too soft to be used and others impossibly heavily pebbled as at Dieppe.

Arrangements could be made to deal with the problem if the attackers could gain the necessary intelligence in good time. To overcome this uncertainty, Combined Operations Pilotage Parties were formed. Under cover of darkness these brave men would travel to the enemy shore in either a small submarine, tiny folding boats or even canoes. They would then bring back ground samples, as well as details of beach exits and other important strategic information.

This reconnaissance was a dangerous proposition, for if they were discovered it could jeopardise the forthcoming operation and possibly get them killed in the process. For this reason, the investigation of the five main beaches on the next Combined Operations target of Operation Torch, the landings on the North Africa coast, were vetoed by the Admiralty. This enormous operation with armies arriving straight from the United States as well as from Britain, was timed to take place about six weeks after the raid on Dieppe.

Like all things military, as the beaches were not inspected prior to the operation a price had to be paid and whilst many landing craft and much materiel were lost in the tidal Atlantic surf, the landings that took place further east in the non-tidal Mediterranean were quite satisfactory.

It was most fortunate that the resistance offered by the French defending troops was only minimal, making this Combined Operation a good training exercise and a resounding tactical success.

At Dieppe the 'bellying' of the tanks in the beach shingle showed the need for a method of allowing tanks to cross difficult uncertain terrain. As a result, another Armoured Fighting Vehicle (AFV) was fitted with a long coir mat or carpet, which was laid out over the beach giving the following tanks a secure foothold. Also, rather than have sappers manually remove beach mines by hand and under heavy fire, another type of AFV was fitted with a massive flail of iron chains that slashed the ground in front of the tank, exploding antipersonnel mines. The flail would clear a 10-foot path over which troops and tanks could safely traverse the beaches.

Based on the Churchill, the Flail tank itself was protected by very heavy armour plate. The Americans ridiculed the use of these 'funnies', completely ignoring the various cogent lessons of Dieppe, and sadly their men had to pay a terrible penalty at Omaha during the D-Day landings.

Apart from the many other lessons in hardware and tactics, Dieppe revealed a chasm in the important and vital area of human relations. Decisions that should have been made on merit were clouded by interservice rivalries, which Mountbatten, at Churchill's prompting, had fought to eradicate from the very day of his appointment to command of Combined Operations.

The problems started early on for Mountbatten. In fact, as soon as he jumped the seniority fence over the heads of some excellent highly qualified more senior officers in the Admiralty. The Admiralty, as also General Montgomery, had not seen the need for another service, Combined Operations Command. Landing operations, they felt, could be dealt with by simply using the Royal Marines.

After Dieppe it became possible to prepare jamming equipment for all the German defensive devices, communications and radar. The plaintive unanswered calls for help by German units and isolated gun posts when denied telephone communications revealed that much of the radio equipment was either Stone Age or certainly low-powered and unstable.

After Dieppe, Mountbatten ordered his radio specialist Lieutenant Commander Robinson's team to prepare the necessary jamming systems for radio, telephone and wireless telegraphy, jamming on the bands as recorded by CPO Dearson and his German-speaking operatives aboard the *Prince Albert*.

For the RAF, the Director General immediately ordered the preparation of what is now called 'white noise' amplitude modulated transmissions (Mandrel) to mask the approach of bomber formations when over the North Sea and on their way to their target cities in Germany.

Shortly after the Dieppe Raid, a primary test was carried out during a bombing raid on Mannheim. This was singularly successful. German-speaking radio telephone operators aboard some of the attacking aircraft heard the panic-stricken response of the German ground controllers when the low-power Mandrel was used operationally for the first time.

Our portion of the Dieppe Raid was really the first ever Electronic Intelligence Gathering mission and it answered many questions, which prior to the raid seemed to require a whole series of probing attacks over a long period.

And time was a commodity in very short supply, as many questions urgently needed an answer.

For instance, how were we ever to be able to place an invasion armada of thousands of ships close inshore without being blown to pieces by the hundreds of radar-directed long-range coastal defence guns? As we had learned to our cost many times, even one surface vessel would be straddled by shell fire if it approached the coast of France in the dead of night.

Another query was how on earth was the navy going to marshal in mid-Channel and in the dead of night the 250 vessels of various speeds, shapes and sizes. The problem was to get this polyglot assembly to approach and enter a tiny mine-swept lane and then fan out to their own beaches in the five different beaching zones.

The enormous degree of navigational precision required could not be achieved using standard naval dead reckoning navigation. The treacherous almost unpredictable tidal currents of the fast-flowing English Channel and perhaps gale-force winds prompted certain experienced naval officers, even those in Combined Operations Command, to predict a complete and utter foul up.

Radio communication of any kind was out of the question as this would have alerted the defenders. A naval operation of this type requiring that degree of precision had never been attempted.

Despite all the gloomy predictions, using his engineering skills coupled with enthusiastic support, Mountbatten came up with the answers.

As mentioned earlier, young Mountbatten had fought to get RAF ASV installed on naval vessels. The *Prince Albert*, the *Princess Beatrix*, the *Invicta* and all the infantry transporting ships were all now fitted with this radar, Naval Type 284, some, of course, with the newly highly secret microwave Naval Type 272. During the February escape of the German battlecruisers from Brest, I described how an airborne type of IFF could be installed on the ground and used as a homing beacon by the Coastal Command aircraft.

For Dieppe, Mountbatten had a seaborne version of the IFF set prepared. It received a seagoing designation of Naval Type 78 beacon. At the time of the Dieppe Raid, however, there was a major problem. The ship's powerful radar transmitter when interrogating the beacon would almost definitely be received by the Freya radar, which would immediately raise the alarm.

Mountbatten's answer was quite cunning and could only have been thought of by a radar engineer conversant with the setting up procedure of the IFF unit. 'Eureka', as these IFF beacons were subsequently called, was a very polite device and was normally completely silent in radio terms. It needed a powerful transmitted radar pulse to bring it into operation. However, the tiny low power beacon, if maladjusted, could be made to 'squitter', that is transmit a steady series of very weak pulses. The ships with their radar transmitters switched off could easily receive these pulses, and by using the direction-finding characteristics of the ASV aerials mounted on the masthead, steer accurately for the beacon.

At Dieppe and despite all the negative gloom mongers, for the first time ever a large fleet of ships under cover of darkness entered a tiny mine-swept lane and, using the Type 78 beacon as a 'fix' or starting point, did a precision run to their drop zones some 10 miles offshore.

From the drop zone, where the Assault Landing Craft were lowered into the sea, they started the long run in to their appointed beaches.

The same marshalling format and five beach landing zones were used equally successfully on D-Day, still more than a year away, and then in the remote distant and most uncertain future.

For D-Day, however, microwave devices not affected by our jamming gave complete secrecy and were to build a very high degree of precision into every facet of the operations.

For the Dieppe Raid and in action for the first time was the new radar, the second version of the naval microwave radar, Naval Type 272. Its extremely short wavelength transmissions could not be detected by the ever-watchful Germans, as they had no equipment capable of receiving these super high frequency transmissions at that time.

Another overdue chore was the urgent need to try under combat conditions a whole series of Combined Operations beach landing naval vessels, all intended to assist in debarquements during battle conditions. Urgent answers on the suitability and serviceability of these unusual craft were needed as our travelling salesman, Lord Louis Mountbatten, was going to the United States to plead for the mass production of huge numbers of these unusual and un-seamanlike, flat-bottomed 'kipper boxes' as they were irreverently dubbed by the Royal Navy.

Combined Operations Command had inherited from Sir Roger Keyes a whole series of ideas for beach landing craft specifically designed for the assistance of troops engaged in proposed beach landings. Mountbatten had inveigled the hard-pressed shipyards of Britain to produce some of these strange vessels, together with his own ideas for assisting troops as they touched down on foreign beaches. These included landing craft that sported two 4-inch guns that could immediately engage the shore defences, so supporting the troops in that critical period when they first hit the beach.

The different vessels all needed operational testing before Mountbatten could finalise the designs and present his requests to his associates in the United States. In the New World he was held in high regard by President Roosevelt, and his friends in Hollywood gave him the media support which he realised was vital to publicise Britain's case before the American public.

Apart from the Assault Landing Craft, which could carry about thirty fully armed men and of which the navy had fewer than half a dozen immediately after Dunkirk, there was a whole series of craft which had been prepared for Combined Operations Command for use in the months ahead. Eventually some 2,000 of the British ALC under the American designation of LCA were subsequently built and used by the Allies in various theatres of the war.

The various types included the LCM or Landing Craft Mechanised, which could carry a tank or motor vehicles and could be used as a beach supply lighter, and the LCP or Landing Craft Personnel, an American-designed wooden craft that could carry about twenty-five men. There was also the LCS or Landing Craft Support, which carried an anti-tank weapon, various machine guns and a smoke-laying mortar, and the LCF or Landing Craft Flat, equipped with two 4-inch guns, Oerlikon and Pom-pom guns and designed to assist the troops during a landing by protecting them from enemy light surface vessels and air attack.

Additionally, there was the LCT or landing craft tank, which could land three heavy tanks or numerous other motor vehicles on a sloping and unhospitable beach.

Finally, at that time there was quite a number of LCIs or Landing Craft Infantry. These could carry perhaps 250 men, and all the necessary assault equipment needed to place them ashore.

All these vessels and others saw service at Dieppe and a vast accumulation of new operational know-how was gleaned under the only circumstances that count in war time – battle experience.

For all beach landing operations, once ashore the soldiers have to be kept supplied. There must be naval fire power on call, air support, logistic support and beach-head organisation to maintain all the above.

Mountbatten realised that rock-solid ship-to-shore communications were a prerequisite and in 1942 suitable small, high-power transmitter receivers that could operate off a beach, perhaps under fire, were not available. He therefore prepared specialised radio equipment that was manned by specially trained men who would carry out all communications as required with a suitably equipped Headquarters Ship to act as a clearing station for all messages that were vital to the operation.

Initially and at Orange Beach these were UHF transceivers 'loaned' by the US navy. The very first Headquarters Ship was being prepared in Scotland. The *Bulolo,* with a large operations room and all the equipment installed in shockproof mounts, was practically completed by June 1942.

The operations room aboard the *Bulolo* was a miniature version of the plotting table at Stanmore, with head and breast telephone communications direct to the plotters. As at Stanmore, an air force controller of each of the various commands, Bomber, Fighter, Ferry etc. was always on duty. With their own sources of information, they would study the General Situation table and take the required action.

On the HQ Ship, the army, navy and air force each had their own controller with totally separate communications to their commanding formations in England and to their Forward Observation Officers who were in the front line with the fighting troops.

Sadly, a strike in the Scottish shipyard delayed the completion of the *Bulolo* and despite Mountbatten's personal intervention, it was not ready for Operation Jubilee. Although a destroyer was not really suited to such specialised work, the *Calpe,* a Hunt class destroyer, was quickly fitted out with all the various types of radio equipment, and the *Fernie* acted as the No 2 HQ ship. As soon as action started, communications between ship and shore broke down with sad results.

Another unfortunate event was that an LCT that was to bring a squadron of tanks to the main Dieppe beach at a most critical moment at the beginning of the landing, arrived instead off Green Beach.

By the time it had made its way to the main Dieppe beach, the fate of the poor sappers who were clearing the beach obstacles was already sealed. When the covering smoke had dispersed, the Germans returned to their guns and tore the young Canadians to pieces.

For the main event, D-Day, 'Eureka' beacons mounted on miniature submarines were to guide the assault forces to the correct beach and at the correct time.

Dieppe and the Combined Operations in the Mediterranean provided the necessary experience and expertise to ensure success on the Normandy landings in 1944.

In October 1942, about six weeks after the raid on Dieppe, came the Combined Operations main event of the year, the invasion of North Africa by large Allied armies from Britain and the United States. The code name was Operation Torch.

The brainchild of Roosevelt and Churchill at about the same time as Dieppe, Operation Torch and the resultant gigantic trans-Africa pincer movement resulted in the ejection of the Italian army, Rommel and his Afrika Korps from North Africa. Huge numbers of Axis personnel remained on the African continent, but now as prisoners of war.

However, Operation Torch, which was supposed to be launched in the first week of October, about six weeks after the Dieppe Raid, was delayed to 8 November, when a huge contingent of American troops landed in Vichy-controlled North Africa.

A rigorous six-month campaign followed with the Allied forces heading east to meet up with Montgomery, who was advancing from Libya. The Axis forces were caught in a gigantic trap. For the first time huge enemy armies were thrown into the prisoner of war camps that were hastily set up all over North Africa.

Almost on the same day that the Germans finally capitulated at Cap Bon, Operation Husky commenced. This was the proposed invasion of Sicily. It promised to be a tough nut to crack. The island had been occupied by German and, of course, native Italian troops, who were soundly established with all the defensive paraphernalia, including Panzer formations and the three types of German radar.

After taking some losses, the Combined Operations Pilotage Parties, now an approved formation, had discovered that most of the suitable coast areas in the southern part of the island had false beaches, some more than 200 yards from the shore. This would cause problems as trucks and other equipment would drown unless a method could be found to bridge the distance from the false beach to the true. Also, as the beaches in the Mediterranean are non-tidal, the usual technique of beaching and withdrawal would not work satisfactorily.

At Dieppe the problem of drowning that had threatened the tanks was overcome by modifying the air intake and exhaust pipes, as well as by introducing waterproofing of the tank body. As a result, some of the Churchill tanks had climbed out of the 8 feet of water without any ill effects.

It was just as well that the COPP parties had sounded the warning, as the same kind of situation might have developed as during Operation Torch, where almost 20 per cent of the landing craft were seriously damaged and many lost.

After Dieppe the importance of surreptitiously inspecting beaches before attempting a landing had been recognised, but shortly afterwards, when Operation Torch was being prepared, the Admiralty, in its wisdom, had vetoed the use of the COPP parties as it might have given warning of the impending landings.

Despite objections by Combined Operations headquarters, the Admiralty issued an edict stating that for Torch, a visual of the beaches by a submarine was all that would be necessary.

In the event it was not sufficient and there were heavy losses in landing craft, transport, men and materiel. It was the loss of landing craft that caused most apprehension at Combined Operations headquarters, as, though they were being produced in large numbers in the United States, most were going to the Pacific.

In mid-1943 there seemed to be no way of obtaining the vast number of beach landing vessels of the various kinds that were going to be required for Operation Overlord, the invasion of France, despite the original modest plan of attack. For this reason, serious consideration had to be given to the problem of the shelving Sicilian beaches as losses here might seriously affect the tally of beach landing vessels of all types available for Overlord.

Combined Operations, however, seemed to thrive on impossible situations and the American navy had mass produced, for another application, floating pontoons of a standard size that could be lashed together to form a floating road. These were brought over to Algiers and to a great extent this solved the problem of getting transport and supplies from the false to the true beaches which abounded around Sicily.

In view of the ongoing shortage of beach landing craft, a way had to be found to retrieve those craft that had been damaged and lay derelict at various places on the North African coast.

The Americans produced a monstrous crane that could be towed into position and used to lift the derelict craft and place them ashore. Using the same principle, a special landing craft recovery vehicle was prepared for future operations where it proved to be of inestimable value. Every operation contributed ever more specialist devices, and all were to be used in the very near future.

For Operation Husky, the landings in Sicily in 1943, the first six Tank Landing Craft Rocket, TLC(R), which had been completed and tested, were brought to the island. It was known that the German defences were well organised, particularly on the critical beaches.

Three of these awful weapons of war were used independently to assist in the fire drenching of some heavily casemated beach batteries. A casemate is a fortified position or chamber, such as a pill box from which guns are fired through embrasures. The other three followed the 51st Highland Division into their assault.

When the 1,700-yard range was reached on the 2-mile range indicator of the specially modified RAF H2S radar, the noisy frightening and murderous barrage started. The Highlanders cheered as they knew that they would need all the help they could get.

The effect on the defending troops of this paralysing bombardment by nearly 3,000 rockets, each with the punch of a 6-inch shell and all in the small area of the gun positions, can scarcely be described. The Jocks landed and took all the gun positions with practically no losses whatsoever. The men of Dieppe had been avenged by weapons born out of Canadian sacrifice.

The Headquarters Ships *Bulolo* and *Largs* had acted as nerve centres for operations on the various beaches and for the first time a fighter Director Ship, albeit put together in haste, was available for air defence. This ship was a freighter with the type 800 mobile GCI RAF radar, exactly like my own one at Bolt Head, mounted on its foredeck. It was a huge success. It proved to be a most wonderful means of defending an invasion force against air attack, not only at that most critical time when hitting the beach, but also for the night and day defence of the whole landing area whilst it was being consolidated. Based on this astonishing success, many more fighter Director Ships were prepared for the next round as well as the main event still about a year away.

Operation Husky, the Sicily landings, was an enormous undertaking with ships sailing from various remote corners of the world to take part in the operation. They came from Norfolk, Virginia in the United States, Kabrit and Shallufa on the Bitter Lakes in Egypt, and Inverary and the lochs of Scotland by way of the Cape of Good Hope, all synchronised to carry out their own part in the assault. As more than 1,200 ships were to be involved, security as to the intended target was hard to maintain.

Operation Fortitude, the plan to delude Hitler on our intentions, had been created. To keep the Germans guessing, it could have been in Norway, the Balkans or continental Europe and Fortitude was always hard at work raising Hitler's blood pressure.

For Husky, members of the Royal Hellenic Air Force based in Egypt kept a high profile to the spies populating Cairo with their overland contacts to Germany via Turkey.

They openly engaged Greek interpreters to assist the new 'secret' project, obviously suggesting to the Germans that Greece and the Balkans would be the next major target for Combined Operations invasion. The Greek Brigade based in the Canal Zone of Egypt believed this to be true.

At that time the writer had taken an 8,000 type GCI out to the Middle East for the defence of the Suez Canal. This was a more up-to-date version of the mobile GCI placed aboard the Fighter Director Ships.

The station was set up at El Gamil, an airstrip 4 miles to the west of Port Said, where we were kept very busy as the Luftwaffe clearly wanted to block the Suez Canal, our major supply route.

For Husky in particular, this waterway was vital as almost a third of the ships would approach the Mediterranean by way of the Suez Canal. In fact, in an attempt to put the Suez Canal out of action, the Luftwaffe had used some very revolutionary aircraft with startling performance.

The Luftwaffe had to carry out continual reconnaissance of the canal, mainly with the objective of catching a slow-moving convoy in the narrow 90-mile waterway and sinking some of the ships. It goes without saying that a ship 'broached to' and immobilised would cause the canal to silt up and be out of action for months.

The Luftwaffe bomber crews were very brave as they had to travel some hundreds of miles from Crete to go about their nefarious work. For the reconnaissance work they used the Junkers JU86P. This was perhaps the first aircraft of that size with a pressurised cabin – even with the latest mark of Spitfire it was almost impossible to climb and manoeuvre at a height of 50,000 feet, the altitude at which these German reconnaissance aircraft operated.

Fortunately, before leaving England my GCI had joined in a secret exercise under the code name 'Exercise Torch' which had been ordered by Air Commodore Tait.

Using one of the newly arrived high-flying B17 aircraft of the United States 8th Air Force, we had discovered some huge gaps in our interception capabilities at extremely high altitudes. The aircraft flew from Land's End to John O'Groats, the full length of Britain, at its maximum altitude and was not even seen by most of the participating GCI radars. As a result, a new amplifying device, now old hat, the 'grounded grid triode' was introduced to improve performance in the gaps high overhead.

We had brought a number of this latest type of mobile GCI, the 8,000 series, out to the Middle East specifically to thwart the Luftwaffe, whom we knew intended to block this important waterway. These GCI radars were fitted with all the latest innovations to help us in our vital task of protecting Port Said and the Suez Canal. With the new equipment we had no difficulty in dealing with these unwelcome intruders and in no time our own prisoner of war camp at El Ballah, located just south of Ismailia on the Suez Canal, was full of Luftwaffe aircrew.

After operations in the European theatre, the interception and shooting down of these luckless brave airmen was like taking money from the blind. We decided that rather than murder these intrepid young lads, we would contrive to shoot them down with our air sea rescue craft in attendance.

El Ballah on the Suez Canal became heavily populated in no time and I know that the young airmen were always addressed with the deference that their bravery warranted.

Many of the ships for Husky sailed from Kabrit on Lake Manzala, where my station was located. Other ships had rounded the Cape of Good Hope on their way to Sicily and the loss of the Suez Canal at that point would have been an unmitigated disaster that would have threatened the successful invasion of Sicily. Despite extreme almost suicidal efforts by the Luftwaffe, the GCI radars of the Canal Zone thwarted their plans, and the ground was now cleared for the assaults on mainland Europe.

CHAPTER 17
ASSAULT AND VICTORY

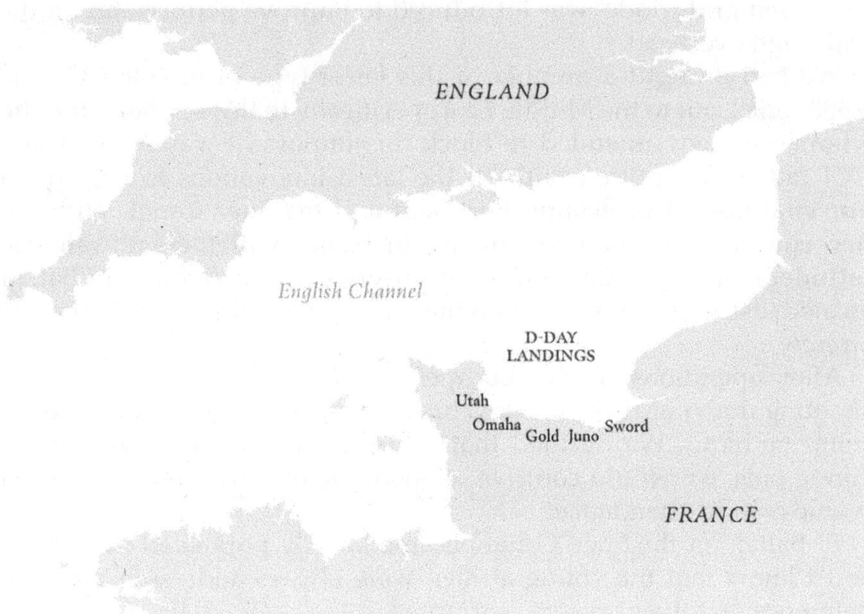

The D-Day landings, 6 June 1944

Combined Operations Command had now come of age. From its inception immediately after the Dunkirk debacle in 1940, when a few completely unsuitable boats tried to show the German victors that we intended to fight, there had been vast expansion of the many types of beach landing craft. Now with much operational experience there was an ever-improving standard of professional skill in the command.

The hard-hitting raid on St Nazaire in 1942 had used ninety craft of various kinds. Later in 1942, nearly 200 vessels of various types had successfully invaded Madagascar in the Indian Ocean, neutralising the French forces and preventing the Japanese from occupying the island.

The Axis powers intended to use its facilities as a submarine base for Japanese and German U-boats.

In August of the same year, in an operation which was a preamble for the D-Day landings, a large naval force of 250 naval craft had contrived a rendezvous out at sea, marshalled themselves in the dead of night, set course and landed on their five allocated beaches in and around Dieppe. They had proved that it could be done.

For Operation Torch, in October of 1942, three massive independent fleets, each coming from different parts of the world, and which included battleships and aircraft carriers, had successfully made landings on three widely separated beaches around the coast of North Africa.

For Operation Husky, in mid-1943, 1,200 ships from three different parts of the world had secretly and successfully taken the island of Sicily, but now, with the main battle on the horizon, General Morgan stated that in his opinion more than 3,000 vessels would be required for the landings in Normandy.

In the event, the expanded plan called for more than 4,000 ships and craft of every kind, not to mention the 6,000 service and supply vessels that were in the tail of the landings.

If one considers the fact that most of these ships had to be conceived, built, modified and mass produced, together with the training of a large force of naval personnel to man the craft, one begins to get a notion of the miracle that had been accomplished by Combined Operations Command in a very short time.

By March of 1943, where key people were being positioned for the next major attack, Operation Husky, the total complement of personnel now in Combined Operations Command was simply colossal.

The headquarters staff in Richmond Terrace in Whitehall numbered 120, with another 6,000 officers of the Command stationed at various establishments around the world, from the lochs of Scotland to the Bitter Lakes of the Egyptian Delta.

There were 45,000 non-commissioned naval personnel, backed up by a huge administrational force of WRENs. In the special service and communications groups there were airmen and soldiers bringing the numbers up to a grand total of almost 55,000. It is hard to believe, but all this had been achieved in just over a year. Just how was it accomplished?

Issuing orders and expecting every man to carry them out just does not work. Leadership from the rear was the norm in the prewar army and navy and this kind of direction would not have achieved this miraculous expansion.

It was the direct personal leadership and personality of Mountbatten, who would climb on a barrel and tell the men to their faces that he expected every one of them to be his own master and make sure that their work, however difficult or dangerous, was carried out satisfactorily and at any cost.

Also – and what was anathema to the British Services – the men were expected to openly criticise and ensure the rectification of observed mistakes.

With a Royal at its head, every man jack of Combined Operations Command felt that he had a personal link with the Royal Family. It was this inspired family style leadership, one that could only be achieved in a democracy, that was to spell doom for the German dictatorship.

For Germany, as with the Allies, radar had proliferated during the period 1940 to 1942. As the operational requirement arose, the German engineers rose to the occasion and produced some magnificent radars on the ground, in the air and at sea.

In 1942, at the time of Dieppe, there were still just the three surveillance radars in general service: the Freya, a long-range precision early warning radar, the type I had visited at Dieppe; the Seetakt, a medium-range naval radar suitable for coastal defence; and the Wurzburg, a precision, short-range gun laying radar with only 12 miles' range on pickup of aircraft.

At the time of Dieppe, we had also discovered that there seemed to be another larger Wurzburg with a possible range of 40 miles on aircraft, though little was known of its actual use. It was surmised to be part of the night fighter control system, as it indeed proved to be.

By 1944 there were some 600 German radars on guard, many deeply casemated with bombproof operating rooms for the radar men. They stretched from Norway down past Holland and France, with numerous installations on the western seaboard of France.

The Freya type had been developed with massive aerials and higher power transmitters giving a very long-range performance, but still capable of precision plotting of ships and aircraft. An extremely clever version of this radar now used electronic sweeping of a very advanced type, which made it into a formidable adversary as it could plot ships on our side of the English Channel.

The Seetakt type had also been improved and though it had a shorter range of pick up than the Freya, it was a super precision gun-laying radar ideal for coastal defence.

With these stations located every few miles along the coast of France, it seemed that getting some 4,000 ships close inshore without detection was utterly insoluble.

There was another family of airborne radars that Germany had produced for the detection of ships at sea. Clearly they were a force to be reckoned with, as they would be able to detect the approach of the invasion armada and sound the alarm.

In the field of communications, the Germans had also scored some victories in their effort to overcome our heavy jamming of the normal radio communications structure. By using the Murray Code, which is virtually a digital signalling system, they had managed to pass messages by teletype, despite all our attempts to corrupt them and jamming.

Our main fear was that they might have also developed microwave radars and communications systems, which were quite impervious to our Mandrel jamming, possibly keeping these pieces of equipment under wraps until we launched our invasion.

They had also produced and used glider bombs with great success on our shipping in the Mediterranean. These would have posed a very serious threat to our assemblage of shipping laying off the coast of France during the landings.

CHAPTER 18
D-DAY - THE POLITICAL SCENE

By 6 June 1944, or D-Day as it is now called, the world situation had undergone a profound change.

The German army had been beaten in North Africa and the vast majority of their men, unlike Britain's disaster at Dunkirk, had been defeated and captured. Even Hitler's Italian ally Mussolini had been vanquished, and Italy had capitulated.

In Russia, the German armies, defeated in open combat, were strategically withdrawing, very much as Napoleon had done many years before, and were retreating back to their Fatherland leaving behind vast numbers of prisoners and most of their arms and equipment.

With the U-boats to all intents and purposes defeated, vast armies had arrived in Great Britain from North America, thankfully with practically no losses.

On the other side of the world, in the Far East, the Americans were retaking the island territories conquered by the Japanese after their stunning victory at Pearl Harbor. Not only had Africa also been cleared of the Germans, but successful Combined Operations had taken place in North Africa and Sicily with the Allied armies advancing up the leg of Italy.

Vast numbers of all types of aircraft had been manufactured in the United States, and the intensive bombing of Germany was repaying the unfortunate German civilians a thousandfold for Guernica, Warsaw, Rotterdam and, of course, the Blitz on Britain.

Eisenhower, still not admitting how wrong he had been in castigating the British for not carrying out the ill-conceived 1942 attacks on the Continent, was now Supreme Commander in Chief of all the Allied forces, which were poised for an assault on Nazi-occupied France.

Rather than invade with the ridiculously tiny ill-equipped force that had been available in 1942, he now insisted that General Morgan's plan for the five-pronged assault over 25 miles of coast be doubled to 50 miles. His 1942 attitude in which he forcefully insisted that the British army had lost the will to fight or at best were scared of attacking the Wehrmacht, had infected much of the American 'top brass'.

In fact, General Bradley's contemptuous refusal to deploy the various armoured Fighting Vehicles and the beach support craft including LCT(G), sporting two 4.7-inch guns, developed as a result of the Dieppe experience, was to lead to a shocking repetition of the 1942 blood bath for the unfortunate American servicemen at Omaha Beach.

It was this basic obstinacy that almost jeopardised the forthcoming landings on the coast of France.

Hitler's problems were increasing. With his Russian armies in headlong retreat, he was unsure of where the Allies would strike in their attempt to regain a foothold on continental Europe.

To keep the Führer guessing, Churchill quietly prepared prospective invasions in various countries to the north and south of 'Festung Europa'.

Would it be in Greece and the Dodecanese Islands using the Greek Brigade, which had been built up ostentatiously in Egypt together with the Royal Hellenic Air Force, also trained and equipped in Egypt?

Another issue that gave Hitler food for thought was the burgeoning armies in North Africa with a massive contingent of Yugoslavian expatriates, many trained by and wearing the uniform of the Royal Air Force, that is, with one small modification to the head dress: the wearing of a red star in place of the RAF cap badge.

With their headquarters at the huge Maison Blanche airport at Algiers, these men kept up a constant air bridge to their brother partisans in occupied Yugoslavia. Transporting arms, equipment and supplies to their Commander in Chief, Marshal Tito, they kept the pressure up on the murderous Nazi oppressors in the Balkans and southeastern Europe.

Setting off from Algiers, large formations of heavy long-distance bombers of the Allied air forces were flying up the Adriatic Sea over Trieste and bombing targets in the Ploesti oil fields in Romania, and even dropping arms and supplies to the partisans in Poland.

The prospect of an invasion of Europe by way of Austria and the south was another of Hitler's nightmares. With Sicily lost and Allied troops approaching Rome, large formations of German troops had to be kept in the south to protect both northern Italy and the Balkans.

This plan to keep Hitler on tenterhooks was known as Operation Fortitude, and it did much to keep Hitler's ever diminishing fighting force spread out over a vast area. In Europe, Hitler had always feared an invasion of Norway, where the Allies would be received with open arms. Since our successful Combined Operations raids earlier in the war, he was convinced that we might one day attack in this area where a lodgement would be easy to attain and yet not be too costly in terms of men and equipment.

By keeping two make-believe armies, one in Northern Ireland and the other in northeast Scotland, visible to German intelligence via supposed lapses in communications security, the Führer was given many a sleepless night. German air reconnaissance in those areas also showed tented army camps, tanks and transport, including landing craft which, though dummies, appeared real when photographed from the air.

Despite Hitler's fears of attack in the far south as well as in the far north, the main possibility was that the Allies would take the short route to Berlin and invade across the English Channel at Calais. For some months the air force had been heavily bombing the mysterious secret weapon sites around Calais and Hitler fervently hoped that the Allies would attempt to invade in this area. They would be allowed to land, and his Panzers would repeat the 1940 nutcracker lesson. To the German generals, all logistical experts steeped in military tactics, this was the obvious choice as the distance of about 20 miles would give the British a quick turnaround for their ships, increasing the speed at which the vast quantities of troops and materiel could be landed.

For this reason, the Pas-de-Calais region was very carefully prepared for the invasion and the subsequent land battle that would bring victory to the Third Reich.

Though we had already chosen our invasion area some 80 miles to the west on the Normandy coast, it was good strategy to foster the German belief that our main invasion force would land at Calais. Under Operation Fortitude, the Allies had put an enormous amount of effort into convincing Hitler that this was our intention.

Tented camps, dummy tanks and vehicles and even troops were used to hoodwink the Germans. There were long lines of make-believe tanks and trucks spread over acres of fields in and around the Cinq Ports of Kent.

Once again two separate make-believe armies used the radio, continually passing all the kinds of information that would constitute the normal radio traffic of two large armies in the field. They even had a commander, the one the Germans seemed to fear most, the bombastic, volatile General Patton. His successes with highly mobile armour in North Africa and Sicily matched the early victories of their own General Guderian. It was Guderian, of course, who had broken the Allied armies in two at the time of Dunkirk and who was mainly responsible for our defeat at that time.

When General Patton appeared in the Kent newspapers walking his dogs in Canterbury, it seemed to verify the fact that the main and most dangerous attack would take place on the short route directly across the English Channel.

If the main invasion were to take place from southern England, both we and the German High Command knew that there were but two options. The attack would either come at Calais or possibly the much longer sea route to the Normandy beaches. It was, in fact, more than four times the distance to Normandy, with the consequent requirement for many times more ships to carry the same payload of troops and equipment.

Hitler and his generals were kept guessing. Would the attack come in Norway, the Balkans, Greece, Calais or Normandy? These were the options and with his men in headlong retreat in Russia, following some horrific defeats, Hitler was very troubled.

The questions were, where would the allies attack and when? As Churchill had planned, the German army was spread over thousands of square miles of mainly hostile territory, and with the Reich under vicious air attack by day from the Americans and by night from the RAF, the Luftwaffe had their work cut out.

One thing was certain, the mighty Panzers, the armoured tank columns that had destroyed the Allied forces at the time of Dunkirk, would be thrown into battle with our invasion forces and this would happen just as soon as it could be arranged. This was a great danger, and the Panzer master Erwin Rommel had already made it known that this was his intention.

Eisenhower had to make provision and prepare to reduce the effectiveness of these battle-hardened men by delaying contact with the invading troops for as long as possible.

The Germans enjoyed interior lines of communication by way of a network of railways and roads, and this would give them a great advantage when competing with ships attempting to disgorge their cargoes possibly under air attack.

The speed of build-up of Allied armour would have to be at least equal to the defending Germans if the invasion was to stand any chance of success. To advise on this cogent problem, towards the end of 1943 Sir Solly Zuckerman, a brilliant scientist, was asked to consider the logistics of this dilemma.

He came up with the 'Transportation' plan which called for a ninety-day 'blitz' on seventy vital road and rail targets, half in Germany and the rest in Belgium and France. They included bridges, railway workshops, rolling stock, railway junctions and marshalling yards. He predicted that if these centres could be heavily bombed, the German High Command would find it very difficult to move their armour laterally across France within a reasonable time frame.

The Panzers could not be everywhere at the same time and Operation Fortitude ensured that the tough tank formations would have to be kept inland waiting to pounce on the bridgehead wherever it was established. The initial operations area in France would obviously be either east or west of the River Seine. If the landing were in Calais, it would be east and if in Normandy, to the west.

Our requirement was to try to trap or encourage the main bodies of German armour to stay to the east of the Seine. This was the preferred battle zone for the Germans, as their tank formations would have plenty of room for manoeuvre and they would be able to repeat their 1940 performance. Their men rotated from the Russian front were tough, experienced warriors with battle proven equipment developed over years in combat with the Russians. It was not surprising that the Allied commanders viewed their first confrontation with the Germans as a rather dangerous test that should be delayed as long as possible.

When the Allied forces first landed, they would be at the mercy of the Germans for the time it would take to land their own armour, troops and weapons.

The terrain around the Normandy beaches was much like the countryside in South Devon, where high, well-established hedges often hid the archaic stone walls that divided tiny farming properties. With line after line of this kind of obstruction, known as 'the bocage', it was certainly not tank country, as both we and the Germans were to learn. Initially for the Allies it became a kind of protective shield, but a shield that was also to inhibit progress into the hinterland of France.

As the problem was to delay the advance of German tanks and troops to the Normandy region, Eisenhower decided to adopt the Zuckerman 'Transportation' plan and destroy all roads, bridges and railways that would help to bring the Germans to the invasion area. The problem was that this 'Interdiction' would require the services of all the Allied air forces, RAF and US, when both were totally involved in their private campaigns against Germany, the RAF by night and the US air force by day.

After some severe defeats at the hands of the German night fighters, Air Chief Marshal 'Bomber' Harris of the RAF had received some unexpected respite when towards the end of 1943, he was instructed, together with the US air forces, to use their bombers to destroy Hitler's secret weapon sites in northern France. Under the code name of Operation Crossbow, heavy bombing of the unusual-looking buildings and earthworks supposedly housing Hitler's secret weapons had been in progress for much of 1943.

Nobody was quite certain what was being achieved by Operation Crossbow and Harris was fuming as he felt that his bombers, with no help from the army, could win the war without resorting to a dangerous and costly invasion of the Continent.

It was a fact, however, that in 1943 his losses on the night attacks over Germany had been so severe that some doubted if there would be sufficient bombers left to carry out his offer to tender victory on a plate.

Under instructions of the Supreme Commander, General Eisenhower, he put a great effort into Operation Crossbow, but obstinately still persisted with some of his aerial bombing of German cities.

Sir Solly Zuckerman emphasised that his 'Transportation' plan would have a vital and decisive effect on the outcome of the coming battles, as his proposal called for the destruction before D-Day of hundreds of specific road, rail and communications targets, which he insisted would inhibit the mobility of the German armed forces.

This was clearly a job for the precision day bombers of General Carl Spaatz and the US Bombardment Groups, who were also totally engaged in their own US daytime bombing campaign, which they also insisted would win the war.

The two Air Commanders, Air Chief Marshal Harris and General Spaatz, after arguing with Eisenhower, finally acquiesced and in the three months preceding D-Day between them dropped 75,000 tons of bombs on the Zuckerman targets.

The net result was that on D-Day, the whole northwest Europe rail and road system was completely dislocated, with more than 75 per cent of the total rolling stock out of action. In the process of bombing the German targets, they also destroyed nearly 5,000 fighter aircraft in combat.

General Spaatz then reverted to his original self-imposed task of destroying German precision engineering factories, as well as obliterating the Nazi oil industry.

These massive destructive day bombing raids deep into the very heart of Germany aroused the same feelings in the hearts of the German fighter pilots as their British counterparts had had in the daytime Battle of Britain in 1940. They realised that they were fighting for their lives as well as the very survival of their country and spared no effort or risk in the horrific day air battles that took place in the skies over Germany.

They were having a wretched time as, since the arrival of the new Merlin-equipped Mustang fighter, they were for the first time meeting aircraft deep inside Germany with a performance far greater than that of their own fighters.

Back in 1942, when the US 8th Air Force carried out its first raids on enemy occupied territory, the crews often reported the shooting down of far more German fighter aircraft than the Luftwaffe had in the air at the time.

In one operation over Brest in which no German fighters made contact with the bomber force, dozens of enemy fighters were still reported as having been shot down. The gunners had been firing at their own fighter escort. I remember a particularly bad incident when returning Spitfire pilots, on landing at my station, Bolt Head, climbed aboard a truck and made for Exeter to confront the US Air Force commanders.

There were tactical problems such as the excessive use by the American pilots of the VHF radio telephones when forming up over England for a raid. These transmissions always alerted the German fighter arm, and they were often waiting for the decoy raids usually carried out by our fighter squadrons to distract attention from the main bomber force. This particular habit persisted and in 1944, 199 Squadron began to fly Mandrel jamming curtains to hide the chatter of the American pilots from the listening Luftwaffe intelligence officers.

In the months preceding D-Day, General Spaatz carried out many heavy and costly daylight raids deep inside Germany in which a large number of important targets were destroyed. As ever, the US air force once again began to make claims of large numbers of enemy fighters shot down. The US intelligence officers queried the enormous figures and disallowed many of the claims, which seemed to be highly exaggerated.

During and after D-Day, when there were practically no German fighters in the sky, it became apparent that General Spaatz and his men had scored an immense and astonishing but unexpected victory, the fruits of which only became evident once our men had landed in France.

There was practically no Luftwaffe fighter arm to interfere with our operations.

Our own aircraft ruled the skies, with the main combat areas now being almost at ground level, where the tank-busting aircraft destroyed any Panzer units and attacked any troops that dared to show themselves in daylight.

With railways and roads destroyed, the German army had to resort to horse-drawn transport in an effort to bring up supplies.

The Interdiction campaign was a huge success and was to make the German supply position in the Normandy bridgehead area a practically impossible task. Despite the heroic resistance of the German forces, they were to be steadily ground down.

With Allied arms, men, transport and weapons being off-loaded from hundreds of ships and a continuous shuttle of hundreds of landing craft bringing tanks and supplies of all kinds, the exhausted, battered Germans would finally be forced to retreat.

From the German viewpoint there was no doubt that the invasion of the European mainland would take place before summer 1944. The exact spot where the landing or landings would occur was a point of anxious conjecture for Hitler and his cohorts. Certainly, our Operation Fortitude suggested that it could be in Norway, the Pas-de-Calais, on the Normandy beaches, on the western seaboard of France or even further east in the Balkans.

A very good guide to our intentions, which could not be disguised, was of course the gigantic build-up of men and materiel in southern England. Hitler and his close advisors, however, always considered that a landing at Calais, where the Channel was only about 20 miles wide, was an obvious choice as the ships would have a short trip and quick turnaround time. In this way a large army could be put ashore quickly. Another plus was that it was the short route to Berlin.

A more difficult proposition was the Normandy coast, where the distances involved were four or five times greater, effectively requiring very much more time to put the same forces ashore.

The Germans have always been great military tacticians, and they considered all the options and came to the conclusion that we would attack perhaps in Normandy as a diversion, with the main punch being delivered later, on the short route to Calais. The Allies pulled out all stops to make Hitler believe that this was the case and that we intended to make the main battleground the hinterland of the Pas-de-Calais.

Much has been written on the German dilemma of that time and in typical style the Wehrmacht, or should we rather say Hitler, decided to keep all his options open. General Erwin Rommel's exhortation to his troops was simple and straightforward: annihilate the invasion force while it was still afloat as this is when it would be most vulnerable. This order could only be carried out if the beach on which the Allies were to land was correctly divined together with the date of the landing.

In each month there are only about four days when a set of conditions of tide and moon would be suitable for a specific landing. One of these dates was 8 May 1944. Intelligence had long since supplied this date and, as a result of nothing taking place, all the German defensive forces on the coast of France were made to stand to.

Nothing happened.

This was principally because Overlord had been delayed one month in order to give the Allied invasion forces in the European theatre another month's production of Tank Landing Craft, which were in seriously short supply. At that date the weather was excellent, but over the next few weeks it progressively deteriorated.

The weather in that part of the world is unpredictable. The best that can be said for it is that it is notoriously unreliable. All four seasons can be experienced in that part of England, sometimes twice a day.

Exactly one month later, the weather had deteriorated so badly that it can only be described as foul. It was a harsh test for the Americans, especially those from the Southern states, who suffered severely from the incessant rain, wind and cold.

Battleships, monitors and specialised vessels were coming from remote harbours and anchorages in northern Scotland, and for this reason some two days' notice was required if the ships were going to keep to their estimated arrival times off the coast of France.

The beginning of June 1944 was a trying time for General Eisenhower, as so much depended on reasonable weather, which seemed to have disappeared for ever.

Southwick House, the SHAEF (Supreme Headquarters Allied Expeditionary Force) headquarters near Portsmouth, was shaken and buffeted by storm-force winds and the driving rain continued unabated. Low cloud and heavy mist in the general Channel area made it obvious to the Germans that an invasion was not on the cards, at least for that moment. It seemed to the German generals that in the raging seas whipped up by gale force winds and ever-recurring squalls that the Allied ships would never be able to keep station and reach their precise positions as per any naval plan. The German navy realised that marker buoys and seaborne navigational aids, so important for the marshalling of what would have to be hundreds or even thousands of ships, would never be able to stay in position under these impossible conditions.

For this reason, General Rommel decided to visit Hitler to try to make him see the light and release some Panzer Divisions for his immediate use in the Normandy region. He had to be capable of effecting a prompt reaction to the landings when they eventually took place.

Both Hitler and Rommel would have been amazed if they had known of a new secret navigational device that made all the station keeping and navigational problems even of a vast armada comparatively simple. For hundreds of years mariners had used the latitude and longitude for navigation. These had now been secretly replaced by a map using electronic coordinates that did not require the sextant and a view of the stars to navigate. With the Decca navigator, the skipper of a ship knew exactly where he was, irrespective of the prevailing light and weather conditions.

The Decca System is a hyperbolic system where the maps do not have completely straight lines but a series of hyperbolae, all easily readable. It consisted of a network of shore-based transmitting stations strategically located around coastlines. Each of these stations emitted low-frequency radio signals known as Decca signals. Each station was unique and by comparing the phase differences between these signals, a receiver on a ship could determine its position.

All naval vessels had their positions allocated months ahead and knew exactly where they had to be and when.

The Germans were fully aware of the RAF hyperbolic navigational system, known as Gee, which they had been jamming for more than a year, reducing its performance deep inside Germany. Gee was essentially a short-range VHF device, and it really depended on a direct line of sight for reception of its signals.

Decca was a very new invention using a very low radio frequency of operation, and being a continuous wave system rather than a radar pulse device, it could not easily be jammed.

All naval positioning of the bombardment battleships, which was so vital to the success of the operation, and of the headquarters ships and the drop zones for the mother ships carrying the troops was based on the Decca navigator.

Also, all minesweeping and marking of the swept zones, together with the placement of the Danbuoys, with their glim lamps, which marked the passage of the ships, was done using Decca.

Furthermore, in the comparatively local area of the English Channel, German jamming of the RAF Gee navigational aid was not effective, allowing it to be used throughout Overlord and all subsequent operations.

One thing is certain: these two navigational aids, Decca and Gee, built an element of precise planning into every single aspect of Operation Overlord. All the other dozens of radar systems that were also used by the Allied forces, and not forgetting those used to confound the enemy, suggest that D-Day was really 'R' day, as without this wonderful invention of radar, Operation Overlord may never have happened, and there would have been no liberation of the peoples of Europe.

Many years have now passed since that great event in naval history, Operation Neptune, the naval portion of Overlord, the landings in Normandy that brought about the liberation of Europe.

Many books have been written on this glorious episode in human history, and all agree that this enormous and dangerous undertaking was the result of great human endeavour, planning and sacrifice.

Various reasons have been broached in many erudite books for the seeming comparative ease with which the Allies brought some 7,000 ships close inshore on that fateful morning, and this despite the watchful eyes of long-range narrow-beam German radar stations, many quite a distance inland and away from the bombardment area.

Admittedly the stations had been subjected to heavy bomb and rocket attack, and many had been put out of action. The point is that even if only a few radar-controlled naval guns had opened up on the armada of ships milling around in the predawn darkness, there would have been an unthinkable disaster.

Many of the naval craft were within 4 miles of the shore and the beaches would have run red with the blood of the soldiers and sailors of the Allied forces. For months the Channel coasts of France would have been littered with the bloated dismembered remains of the brave invaders.

The 1942 Dieppe Raid had rammed home two very basic hard truths on proposed landings on enemy held territory: there was no point trying to capture a port intact as it would be destroyed in the process; and, most vitally, the defending troops must be virtually blasted and bombed out of existence or at the very least completely disorientated before the actual landings take place.

Since time immemorial the most effective method of reducing a land-based fort was by a nerve-shattering naval bombardment.

A battleship capable of despatching huge shells the size of a truck loaded with high explosive can plough up the ground like no other projectile. Admittedly an aircraft dropping 10,000-lb and 20,000-lb bombs can also be a nuisance to the defending troops, but a steady two-hour bombardment by battleships with their fire directed by a Forward Observation officer is without doubt the most effective way of convincing an enemy that he is on the wrong track.

If Operation Overlord were to stand any chance of success, the battleships and the invasion flotillas would have to take up their allotted bombarding positions in the predawn darkness with no warning whatsoever to the ever-vigilant defending German gunners and troops.

It was a tall order, and when Rear Admiral Alan Kirk of the US navy and Rear Admiral Sir Philip Vian of the Royal Navy were informed secretly by Air Marshal Leigh-Mallory that on that particular night the German early warning radars would be completely out of action, they expressed their disbelief in no uncertain manner.

Nevertheless, when the British Admiral had, under specific orders from his superior, Admiral Ramsay, finally agreed to instruct his battleships, cruisers and monitors to take up station just a few miles off shore, he, in a very loud aside was heard to say, 'It was all a lot of nonsense.' From his viewpoint and considering the rough handling the navy had endured for nearly four years at the hands of German radar-directed coastal guns, he had every reason to fear for the safety of his capital ships and their crews.

Fifty years on, in their writings, these brave and talented seamen still demonstrate very clearly that they cannot visualise the technical miracles that were accomplished on their behalf and, for that matter, for all humanity.

One thing was certain. All the landings would only have a fighting chance of success if the battleships of the five bombardment groups could slickly take up their positions about ten minutes before dawn without being detected.

If they were picked up on one of the many German radars still functioning, Rommel's order would have been obeyed: 'The enemy must be annihilated while still at sea…'

If the German gunners had fired blindly into the darkness, the battleships and the thousands of supply ships compressed just offshore would certainly have been disrupted and perhaps decimated. As Churchill always feared, Operation Overlord would have become a naval disaster of the greatest magnitude, a defeat from which the Allies might never have recovered.

To Admiral Vian and his team of experienced and brave seamen, the thought of ever being able to get close inshore without a shot being fired simply did not make sense. They had no idea and could not be told the detail of how the Allied radar men had prepared a complex and detailed plan to completely deceive and mislead those German naval officers closely watching the operations room table at naval headquarters on that fateful night. The German coastal defence guns our 'intelligence' organisation had identified, each with a tried and proven radar fire control system, were more than capable of disposing of the huge compact marine target which was to gather a few miles offshore. The Allied admirals simply did not believe it possible that the excellent German gun-control radars at the hands of which, over the years, our naval vessels had taken severe punishment, would by some magical means be put out of action.

The Neptune plan called for some of the larger bombardment ships to station themselves less than 6 miles offshore and the Americans simply would not comply as some of the guns, for example those at Îles Saint-Marcouf and Pointe du Hoc, supposedly had a useful range of 15 and more miles. For this reason, they decided to take up station a full 12 miles from the shore; when the batteries in both these areas were finally captured there were no guns in position.

Having sampled a 10-mile, two-hour night trip on a Landing Craft Assault, I have an inkling of what it was going to be like for the US lads, especially if the weather was uncooperative.

The naval commanders' feelings, US and British, can be easily understood, for if the German radar was not put out of action the battleships, even in darkness, would have been severely damaged and more than likely destroyed.

In the event, the British ships obeyed instructions, and many took up their positions less than 6 miles offshore, but the US ships just less than 12 miles.

As portrayed so magnificently in the film *The Longest Day*, the German gunners had no idea that the invasion fleet had arrived until the dawn mist lifted and the naval bombardment began. The landings did achieve surprise, and in the many books that described that miracle, various reasons have been advanced for the 'unbelievably good fortune'.

This, of course, is a simplistic assessment of an operation in which, as we will see, dozens of different radar systems on ships and in aircraft were used together with carefully planned countermeasures, decoys and other clandestine activities, which had over a long period of time been specially prepared for this party.

Perhaps we should zero in on the critical issue of what the defending German High Command thought on the night of the invasion. This, of course, with hindsight we can do quite well.

Once the noisy dummies dressed like paratroopers and firing crackers began to land in the very early hours of the morning, it became obvious that we were in the midst of an effort to delude the High Command. The wily Germans decided not to be fooled by our tactics and watched and waited for the seaborne invasion, if indeed it was an invasion and not just another scare. German Seetakt naval radars and specially despatched ASV-equipped aircraft reported glimpses of an armada of ships approaching the Dieppe region.

In view of our 1942 attack in the same area, this seemed to be a reasonable and probable target. All the many plots placed on the table could be clearly seen. That is until heavy jamming obscured the approaching fleet. From time to time more plots were tabled despite intense jamming.

It all made perfect sense.

They were, of course, plotting Operation Taxable, which seemed to be a huge armada of ships approaching the coast of France at the correct speed for a convoy of that size.

The jamming was deliberately reduced at times to allow the radar to follow this massive 'armada' to ensure that it was plotted on the German operations room table for all to see. But then another fleet of ships farther to the east was detected and this group was approaching Boulogne. This too made sense as this is where the Germans were convinced that the main attack would finally have to come. The first fleet, it seemed, was a feint to mask the new track, which would be the main attack.

This armada came to a halt some 8 miles offshore and with the aid of loudspeakers made all the noises that might be associated with the lowering of landing craft and the marshalling of ships and men. This fake armada was codenamed 'Glimmer'.

Definite plots from these two regions were anxiously followed by the German High Command and there was nothing else on the table. From the Normandy coast there were no plots of ships and shipping, and the table was, as was normal, absolutely clear in that area.

This is not surprising as starting from midnight, a curtain of 'mush', a dense, indistinct mass of signals, had slowly affected those screens of those early warning stations in and around the Normandy coast; that is, of course, those stations that were still operational.

This was not surprising as 200 ships fitted with Mandrel jammers were slowly approaching the coast of France.

To the German radar operators, the 'mush' seemed of little consequence as initially ships and aircraft could still have been plotted at reasonable ranges. There seemed to be no urgency as the gusting winds and driving rain indicated that this was not invasion weather.

It must also be remembered that for more than a month, all German radar stations had been subjected to attack after attack, and many were working under conditions of reduced performance with high 'noise' the result of damaged aerials and temporary repairs.

The 'mush' observed on the screen would be reported to the duty mechanic when he came on duty. He would then replace the front-end tube or check over and repair the aerial system once again. By 7 am, when the duty crew was to be changed, the mechanic's services were no longer required.

It was a replay of the events of 12 February 1942. In those critical three hours when the RAF crew had returned to St Eval thinking that their ASV II was faulty, the *Scharnhorst, Gneisenau* and *Prinz Eugen* had escaped from Brest.

On this occasion it was the German radar organisation that was duped.

With the success of the D-Day landings, Air Commodore Victor Tait had avenged the ignominious defeat inflicted on RAF 60 Group in 1942.

Now, as Air Vice Marshal Sir Victor Tait, he continued his magnificent work as Director General of Signals for the RAF.

CHAPTER 19
OPERATION NEPTUNE

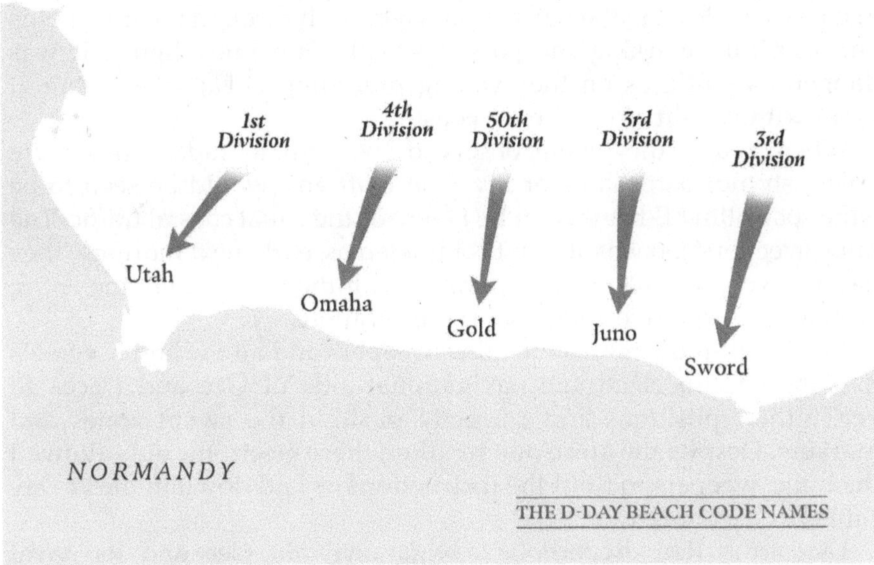

Operation Neptune, the naval portion of Overlord, in which the equivalent of a city the size of Birmingham with all its utilities was to be transported bodily to France, was, of course, a stupendous, hair-raising and danger-ridden exercise. This was only one part of an operation that was to include separate large landings by glider and parachute troops dropped well inland to save bridges and other strategic communications routes from destruction by the retreating Germans.

British airborne troops were to carry out these tasks inland from the British landings and US paratroopers were to be dropped to prepare the way for the US seaborne troops landing on Utah Beach. The whole seaborne attack was divided into two, US and British. There was still the vestige of the Combined Operations planning in that there were to be five separate landings, as at Dieppe, but they were now to be spread over 50 miles of coastline and not 10, as in 1942.

The three landings to the east were British and Canadian, and the two to the west were American. The most easterly landing was Sword Beach, followed by Juno and Gold.; these were the combined Canadian and British operations. The Americans were next at Omaha Beach, and the most westerly, at the base of the Cotentin Peninsula, was Utah Beach.

Starting from the previous day, minesweepers had been working like beavers clearing the narrow passage south through the minefields and carefully marking the passages with Dun buoy lights. It was dangerous work, as on the evening preceding D-Day they were at times within sight of the French coast.

When and if they were observed by German radar, they were obviously not battleships or invasion craft and would be seen to be either patrolling E-boats from Le Havre or the usual coastal traffic. The minesweepers' job was still not completed as, early next morning, they had to sweep the whole area parallel with the coast where the entire armada was to arrive and take up their stations.

In all their movements the minesweepers and all the larger vessels had the superb electronic navigational aids of Gee and Decca to verify their positions and correctly position the swept zones and markers. Despite the atrocious weather, these electronic aids allowed the minesweepers to fulfil the instructions as laid down in the D-Day 'Bible'.

Using the RAF hyperbolic navigational aid Gee and its naval equivalent Decca, both of which supplied a completely new set of coordinates entirely replacing latitude and longitude, they could, despite the shocking weather, work with utter confidence and an incredible degree of accuracy. Their work was vital as they had to clearly mark the mine-swept waterways to be used by the thousands of ships carrying the assault force and its supplies and services.

During the Dieppe Raid, those of us landing on Green Beach were placed on the beach in a wrong and most inconvenient position – we were dropped in front of the houses billeting the local German defence garrison. To avoid this kind of fatal mistake from happening again, Combined Operations had placed two tiny submarines just off the beaches to mark the exact point of touch down, as well as the limits of the British landing area. At the appointed time they were to surface and put up an 11-foot mast with red and green flashing lights, as well as a corner aerial reflector and a Eureka radar transponder.

In this way, and despite the darkness, the skippers of the pilotage craft could read their range in miles to the beach, and the landing craft following could steer by the coloured lights to make landfall at exactly the right spot.

As these coloured lights might not be seen from a distance, at the last moment the submarines had to go further inshore, running almost aground, and put up a powerful white light. This would penetrate any predawn sea mist that might be present and give the landing craft skippers a long-distance general visible marker beacon.

These brave men spent nearly three days virtually asphyxiated in their tiny craft within visible distance of their markers: a tiny church on one beach and a small hill near the mouth of the River Orne. Had they been seen, they would undoubtedly have been killed and, judging by my own sad experience at Dieppe, the D-Day landings might easily have failed miserably with unthinkable loss of life.

The airborne soldiers were also in extreme danger as the glider men had to make the trip across the English Channel at low altitude in slow-moving, towed gliders, and the paratroopers in equally slow air transport aircraft.

The German night fighters of that period were past masters in the art of aircraft destruction at night. If they were directed in the general location of the massive armadas of airborne troops, they would literally have made mincemeat of the defenceless flying troops. The carnage and destruction would have been dire and, furthermore, the men all had vital tasks to perform such as preventing the retreating Germans from destroying strategic bridges and securing the passages across causeways in inundated countryside. If the Germans had succeeded in attacking these brave men, it would have seriously delayed the general advance and progress of the seaborne troops once they had landed.

To ensure that the airborne fleets were not molested, the RAF organised a huge radar and radio countermeasure campaign using techniques developed during the bombing raids on Germany.

First of all, as many as possible of the German night fighters would have to be lured to a supposed prime target miles to the east of the Normandy landings. This was accomplished by Lancaster and Flying Fortress bombers of 101 and 204 Squadrons, which generated a 'spoof' bomber stream by flying south into France, following the general track of the River Somme. Bombing raids in this area had been carried out many times in the past and the German night defence organisation swung into operation. The Luftwaffe night fighters were scrambled and entered the stream, beginning to attack the bombers on their way to their supposed target. The bombers, heavily loaded with 'Window', poured great quantities of the tinfoil into the night sky, simulating a raid by hundreds of bomber aircraft. The German ground controllers, fearing another massive raid on an as yet unknown target town, threw every fighter they could muster into the fray.

Familiar with the RAF preparations for a major bombing attack, more night fighters were scrambled and directed to the area of the raid when it was marked by the customary coloured flares. In the early hours of the morning, when the bombing started, a major portion of the Luftwaffe fighting force had arrived on the scene thirsting for blood.

At that point, and coordinated with the arrival of the airborne troops over France, the aircraft of 100 Group, the radar and radio countermeasures group of the RAF switched on a massive barrage of communication-jamming devices. Some of these were 'Jostle', 'Air Borne Cigar' and 'Tinsel'. These were airborne radio telephone jamming transmitters of huge power that wiped out all ground communications to the night fighters, trapping them in the vicinity of the 'spoof' raid. At exactly the same time, a barrage of other land-based jammers in England opened up, rendering all radio telephone communication to the German night fighters utterly impossible.

This countermeasure was not unusual, and the German night fighter pilots, using their own Lichtenstein SN2 airborne AI radars, were, once they knew the direction of the bomber stream and the supposed target area, more than capable of locating bomber targets without any help from the ground.

If, as could easily happen, the airborne troop transports were detected by inland radars, they would still be reasonably safe as the available night fighter force was totally engaged and incommunicado many miles east of the drop zones.

This, however, was not sufficient protection for the airborne troops, as any radars involved in the plotting of aircraft over western Germany might quite accidently get a glimpse of the masses of aircraft in the Normandy skies. Night fighters from Eastern German bases would then have been ordered to France to intercept and destroy them. To avoid this possibility and to put any surviving Channel-based German radars out of action, a huge airborne Mandrel screen was flown for the first time. This was to prevent pick up of the air fleets leaving England by the superb German long-range early warning radars such as the Mammut and the Wasserman, which had been assembled by the Germans. These stations were extraordinary improvements on the standard Freya AN radars, such as the one I had visited at Dieppe in 1942. Thankfully, they used the same long wavelength of operation and were susceptible to our mass barrage of airborne Mandrel jamming.

The airborne Mandrel screen that was maintained from D-Day onwards was set up as a kind of impenetrable curtain from Portland Bill to the east and then north up to the Wash. It ran north drawing a line 80 miles out to sea off the coast of Holland and prevented any further long-range pick up by German radars of ships and coastal traffic, now deeply involved in Operation Neptune and any subsequent operations. This Mandrel curtain, flown by 199 Squadron at 18,000 feet, supplemented the 200 multi-kilowatt shipborne Mandrels that came to life for the first time when General Eisenhower said, 'Let's go.'

As a result of this stupendous synchronised jamming effort, with one tiny exception, no attacks were carried out on any aircraft or ship during the critical first night of the D-Day landings.

The whole story should also include the aforementioned two feint 'invasions', Operations Taxable and Glimmer. These were both part of Operation Fortitude, the scheme to keep Hitler guessing as to just where the invasion would take place. The German High Command believed and expected our main assault would have to take place on the short Channel route to Calais. A carefully rehearsed, dangerous and difficult operation was planned to confirm this belief and delude the German radar into proving to the German High Command that their assumption was correct.

Using electronic subterfuge, it was decided to indicate that a huge fleet of ships was approaching Boulogne in the early hours of D-Day. The plan called for a long stint of dangerously low flying at night and an extraordinary degree of navigational skill by a large number of bomber aircraft. Who better than the proven experts in low flying at night, the boys of 617 Squadron, the Dambusters.

The idea was to fly the aircraft in a series of elongated elliptical tracks, very much like a 10-mile-long racecourse. The 'racecourse' had to advance from England towards Boulogne at the speed of a convoy of ships, less than 10 miles an hour. If the aircraft had been observed by the German radar, they would quickly deduce what was taking place. The blips of the aircraft had to be blurred and made to beat slowly, in order to look like shipping.

The blip of signal seen on a radar screen from a large ship is distinctive and easily recognisable, so the plan called for some small surface craft, each flying and towing a captive balloon, to move towards France directly under the racecourse track being flown by the Dambusters.

These 30-foot balloons were known as Filberts, named after a well-known cartoon character of those times. Each Filbert had a 10-foot metal radar reflector built inside the fabric envelope. It gave an enormous slowly beating blip on the radar screen and could not be identified as anything other than a large ship. But here was the rub: only one ship.

To improve the 'spoof', about six launches were used and even these might have been seen at times to be separate surface craft. To fudge the picture, the low-flying aircraft would have to follow the launches flying their racecourse tracks and slowly approach the coast of France.

There was still the odd chance that an astute German radar operator would unravel the strange picture of aircraft and ships. To really complete the job of blurring the radar picture and make it really appear to be a fleet of ships, the aircraft were to dispense 'Window': the slowly falling pieces of tinfoil that would put the finishing touch to the deception.

Other aircraft of 199 Squadron flew near the Glimmer spoof, switching their Mandrel jammers on from time to time to make the observed picture more difficult to read and seem that we were trying to mask our invasion convoy.

When the make-believe invasion fleet came to a halt 10 miles off the coast, they used recorded seafaring noises which were amplified and broadcast towards the shore to convince anyone listening that a fleet was out there disgorging its landing craft and troops.

Another operation, codenamed Taxable, exactly the same as Glimmer, approached the French coast slightly to the west of Dieppe. Of course, this operation, with its need for incredibly accurate flying, could not have been carried out without the RAF navigational aids, Gee and G-H, which were medium range radar navigation aids introduced in 1942.

One or two surviving German radars did, in fact, pick up signals and another as yet unmentioned team on motor launches fooled them by recognising their signals and replying to them with another radar countermeasures device first used two days before the Dieppe operation in 1942. This was codenamed Moonshine, which could only fool one station at a time and using only electronic means could be made to look like a fleet of ships.

There were three separate Moonshine interrogations, and each was handled excellently, giving the impression that a fleet of ships was approaching Dieppe.

Moonshine was first used back in 1942 to indicate that a large daylight bombing attack was to be made on the Renault motor works in eastern France. The German fighters took off to intercept the ghost aircraft, and this kept them well to the east when Flying Fortresses of the US 8th Air Force overflew the German airfield at Abbeville on their first operational mission over Europe. This was to bomb Rouen railway junction two days before the raid on Dieppe.

This was the US 8th Air Forces first venture into German air space, and we were all hoping that this operational flight would not result in heavy losses to Adolf Galland and his Abbeville boys. Due to this first Moonshine 'spoof' there were none, but at 10 am on the morning of 19 August, when we were resting at the roadside on the road to Petite Appeville, I saw and heard these wonderful high flying aircraft bomb Abbeville, the station that was so bothersome to me and my RAF station back home, Bolt Head.

The net result of all these closely integrated counter measures was that the ships and airborne troops of Operation Overlord arrived at their targets unmolested - that is apart from the aircraft carrying the 82nd and 101 US Airborne Divisions, who chose to ignore our pleas for routing the aircraft through the Mandrel-jammed air space.

The drop zone was apparently changed because a large German force had been reported in the area. Instead, they approached from the west, overflying the Cotentin Peninsula where, as they had been warned, they were subjected to an intense anti-aircraft gun barrage that distributed the Americans all over western France. The few of these men who did reach the target area, however, did succeed in carrying out their allotted tasks.

The rest of the troops were scattered all over Brittany and this more than anything else paralysed the German defence. They could not prepare a counterattack as they could not locate the focus of the attack. It was a case of the Germans not being able to find out what we were up to, as we did not know ourselves.

Mountbatten always referred to this very real phenomenon in warfare as 'the fog of war'. As ever it was the quality of the men and their training that decided the issue. The few who found the causeway exits secured them for their seaborne friends and the troops poured inland.

Utah became a huge success with very low casualties.

Sword

Sword was the most easterly of the landings and dangerously near the long-range, heavy calibre guns of the port of le Havre.

It was a British landing and on the face of it the hardest and most dangerous nut to crack.

Three fearsome 16-inch guns just north of the harbour of Le Havre were known to be manned by skilful crews and the huge shells had a range of 25 miles.

Hopefully, the guns of HMS *Ramillies*, *Warspite* and the gun platform *Roberts* would give the German gunners more than a run for their money.

Also included in the bombardment team were the cruisers *Frobisher, Arethusa, Mauritius* and *Danae,* together with twelve destroyers, all of which quietly approached their firing positions under the protective cover of Mandrel jamming.

The heavy bombardment ships made their way at speed through the swept mine lanes to their positions (as indicated by their Decca navigators), anchored and opened fire immediately.

Besides the menacing 16-inch German guns, it was known that there were three 12-inch naval guns, thirty-five 6-inch guns, four 3.9-inch guns and a battery of 6-inch guns a mile or so inland. To match this formidable array of weaponry the navy had pitted its best.

Most of the large German guns had proven expert radar control and the risks bringing Britain's largest battleships to these extremely hazardous positions were very high. The 16-inch Le Havre guns had a range of 25 miles and when Vice Admiral Patterson had been ordered to station his ship just a few miles away from this massive array of land-based heavy weaponry, he can be excused for viewing the prospect with horror and trepidation.

However, he had his orders and stuck to them, despite his forcibly expressed doubts on Air Marshal Tedder's claim that the German radar would not work at dawn on 6 June. In true British naval tradition, he obeyed his orders and got on with the job. He had good reason to worry: the Royal Navy had suffered years of chastisement at the hands – or barrels – of these very guns.

In the event, the RAF promise was fulfilled. Also, it turned out that heavy RAF bombing had put much of the enemy fire power out of action. This, of course, was not known at the time, but the ships bravely raced into position, took up their bombardment stations and went into immediate action.

The hundreds of ships and specialised landing craft of which the Sword landing force was composed were led into battle by one of Mountbatten's Combined Operations HQ ships, HMS *Largs,* a far cry from Mountbatten's original HQ ship at Malta in 1935.

The beach landing support force included five rocket launching landing craft packing more than a thousand shells, each missile with the punch of a 6-inch shell, three heavy gun landing craft, five flat landing craft, eight regiments of SP (special purpose) artillery, and eight armoured tank landing craft together with a mass of other support craft making a total of 285 ships. The magnitude of the naval assault force beggars simple description and this was just one of the five phalanxes of naval vessels with which the Germans were going to have to contend on that historic day.

The Sword attack was opened by forty DD (duplex drive) tanks (a type of amphibious swimming tank) that could swim the last few yards into the beach, and this operation was carried out under a smoke shroud to make the German gunners' job a trifle more difficult.

The weather and seas were shocking, and many DD tanks did not survive, but those that did were guided to the correct touch down point by the midget submarine *HMS X23* that had been waiting not so patiently for more than two days.

The troops followed, together with the Flail Tanks, the Bobbins, the Crocodiles and the other special armoured fighting vehicles developed at the Combined Operations Experimental Establishment in North Devon.

Unlike Dieppe, the troops did not have to pit flesh and bone against the enormous blast of the German artillery and mortars. It was not a model Combined Operations exercise, as the losses were still high, but fully aware of their tasks, the men used their specialised fighting vehicles to great effect.

The submarine guided the DD swimming tanks to the touch down point right on time. Under their cover, and some minutes later, the armoured fighting vehicles brought and protected the engineers who removed many of the obstacles, allowing the flails to beat a path to the German defenders, who were quickly dealt with by the various other AFVs designed after Dieppe for precisely this kind of action.

Theirs was a tough battle, but with the aids provided, the troops got off the beach and made their way inland and thankfully relieved the paratroopers of the 8th Airborne Division, who had seized the bridge over the River Orne during the early hours of that morning.

At one point some German E-boats appeared out of the smoke but did little damage with their torpedoes, though sadly they sank a Norwegian destroyer, the *Svenner*. As should be expected, there were casualties, but these were far, far fewer than what had been expected.

All in all, Sword was a magnificent triumph, which the men and the Combined Operations planners had worked for and richly deserved.

Juno

The landings west of, and next to, Sword were codenamed Juno. There was just one gun battery facing Juno and it was composed of four 3.9-inch guns. However, all along the waterfront, Rommel had installed heavily concreted small-gun emplacements that fortunately were manned by east Europeans who were no longer enchanted by Hitler's war.

The HQ ship HMS *Hilary* led the force to the lowering position only 7 miles off the beach and the bombardment force, the cruisers HMS *Belfast* and *Diadem* and eleven destroyers took up their positions and opened up on the enemy emplacements. To support the troops there were eight rocket landing craft, four regiments of special purpose artillery, seven landing craft with large guns, six flak landing craft, six tank landing craft with high explosives and a multitude of other support craft totalling just under 190 ships.

The heavy seas had slightly displaced some of the marker buoys which delayed one of the craft from reaching the start line on time. With the Combined Operations control launches fussing around the ships in the uncooperative sea, and after a considerable delay, the ships were lined up like race horses at the Derby. Rear Admiral Dalrymple-Hamilton then gave the order to go, and the rocket craft released their blood curdling, shrieking salvo of rockets, completely demoralising the defending troops.

The delay caused trouble as the beach obstacles to be demolished were covered by the rising tide, but the armoured fighting vehicles gave the men cover as they cleared the way for the specialised beach fighting force and the Canadians waiting to get to grips with the casemated defenders.

After a demonstration of the Crocodile, a flame-throwing tank, many of the highly motivated troops ignored their German NCOs and stopped fighting. There were still pockets of resistance, but the Canadians, making full use of the various armoured fighting vehicles, did a splendid job, and the beach was cleared with remarkably few casualties.

The submarine *HMS X20* that had guided the force into the correct landing spot left the beach and was towed in the uncomfortably heavy seas all the way back to Portsmouth after a very tiring and vitally important task had been accomplished.

The marking of the correct point of touchdown was terribly important as I had learned only too well at Green Beach during the Dieppe Raid: it could make the difference between immediate success and absolute disaster.

Churchill's adage on 'The Few' should equally be applied to this tiny group of gallant submariners who ensured our success at these two vital landings, Juno and Sword.

Gold

The landings on Gold were the most westerly of the three Canadian and British landings. The guns to oppose this landing were four 6-inch, one 3.4-inch, two 2.9-inch and four 4.5-inch.

The HQ ship for Gold was HMS *Bulolo*. This, we may recall, was the HQ ship that Mountbatten had prepared for the 1942 raid on Dieppe, but it had been delayed by labour troubles. On D-Day it reached its lowering zone only 6 miles from the shore without any problems.

Leading the attack force were the bombardment ships, the cruisers HMS *Ajax, Argonaut, Orion* and *Emerald,* together with the *Flores* and fourteen destroyers. Strong resistance was expected so the bombardment group opened up as soon as they had dropped anchor.

The support craft for the troops were eight rocket-launching craft to blitz the beach defences, seven flat landing craft, four large gun boat landing craft, four large support landing craft, three regiments of special purpose artillery in sixteen tank landing craft, together with the rest of the supply and support numbering 243 ships.

The softening up bombardment started at 5.45 am and lasted a full two hours, putting many but not all of the opposing guns out of action.

Recognising the importance of the HQ ship, the German gunners forced her to change station by a series of near misses. The *Ajax* replied with some precision shooting and quickly put this gun emplacement near Longues out of action.

The timing of the whole landing went like clockwork, very much as if it were a Combined Operations exercise, and this despite the lack of cooperation from the weather. As the conditions were very bad, all were thankful that the troops only had to endure a 7-mile trip into the beaches. Even so, they were cold and seasick and glad to set foot on shore.

There was no requirement for submarine markers or a precision landing, but the troops were faced with a series of difficult tasks, some of which were allotted to Commando troops specially trained to assault the specific objectives.

The landing was eminently successful, and the rest was up to the army, who had a difficult slogging march ahead.

Utah

This was the most westerly of the landings and formed about a third of the United States' part of Operation Neptune.

The US navy, army and airborne troops were to carry out the landings under a plan designed and implemented by their own staff, and it was to be a practically all-American operation.

It was to take place at the base of the Cotentin Peninsula, which was protected to a great extent by the land mass to the west. The peninsula is a long narrow stretch of territory that reaches out into the western boundary of the English Channel, in fact it is western France bordering on the Atlantic Ocean and the Bay of Biscay.

Completely disregarding the unpopular Air Marshal Leigh-Mallory's instruction to route the aircraft carrying the US airborne troops on a southerly course that would take them over the western side of the Utah landings, the Americans made their own flight plan.

This was to prove disastrous.

The reason for the dislike of Leigh-Mallory by almost all the senior commanders was the result of a series of sad events that followed the Battle of Britain in autumn 1940.

In concert with a junior officer, Wing Commander Douglas Bader, Leigh-Mallory had carried out an insidious campaign and secured the dismissal of their own senior officer, the chief of Fighter Command Air Chief Marshal Hugh Dowding. They alleged that by using faulty tactics, he had almost lost the Battle of Britain.

Not only had Dowding been 'retired' by Churchill, but also his deputy and the man who in truth successfully stage-managed the defeat of the Luftwaffe, Air Vice Marshal Keith Park, the Air Officer Commanding 11 Group.

Of course, Dowding's conduct of that historic battle was governed entirely by his skilful use of 'Enigma' intercepts, the existence of which could not be revealed to his junior officers. Churchill should have come to Dowding's defence, but for his own reasons, he did not.

We must remember that starting from 1936, Dowding had fathered the whole wonderful and successful defensive machine that was Fighter Command and was subsequently knighted.

For this reason, which was now well known, Operation Overlord Air Force Chief, Leigh-Mallory, was held in poor regard by all.

They decided to ignore his orders and routed the low, slow flying transport aircraft over the Cotentin Peninsula, where they had to run the gauntlet of dozens of Wurzburg-controlled anti-aircraft batteries that were in no way affected by the Mandrel jamming. The mass barrage of precision anti-aircraft batteries blew dozens of aircraft to pieces and scattered the rest all over northwest France.

Leigh-Mallory was now in the same position as Dowding had been in 1940. He now also had a secret that he could not reveal to anyone: the proposed sudden use of super high-power Mandrel jamming of the German early warning radars, those that could spy on the approaching armada.

The US navy also discounted the secret information that the German early warning and gunnery radars would not work on that night. Because of the threat they felt was posed by radar controlled coastal artillery, the US navy had decided that their ships would anchor at a transport zone far out to sea. In fact, a full 12 miles offshore.

The flagship was the USS *Bayfield,* and the bombardment force consisted of the battleship USS *Nevada,* the British monitor HMS *Terror,* the cruisers USS *Tuscaloosa* and *Quincy* and the British cruisers *Black Prince, Hawkins* and *Enterprise,* together with eight US destroyers and the HNLMS *Soemba.*

The US force had only 116 assault ships, and as there had been a false start, only 130 of the 260 non-combatant support craft arrived later that afternoon.

The 9-mile stretch where the landing was to take place was mainly sand beach, in part backed by about a mile of sand dunes. Behind the beach the land had been flooded and access to the hinterland was mainly by way of four narrow causeways from the sand dunes.

The Todt Organisation had built a long, low wall, and the Germans had installed thirty fixed and mobile gun batteries of low calibre, intended mainly for beach defence and totalling over 100 guns of various types.

This was in line with the Rommel plan to stop the invaders when they were most vulnerable, whilst they were still afloat.

As the troop transport ships were far out from the shore, the assault landing craft had to weather very rough seas, and the troops suffered severely as the craft fighting a headwind struggled to make the two-hour journey to the beach.

The sea trip was so long that pilotage patrol launches escorted the landing craft as they had no adequate means of navigation and would almost certainly have lost their way in the strong, tidal waters.

It was a rough ride, with the craft taking on water and most of the men exhausted and seasick.

As they approached the shore conditions improved when they were protected somewhat from the wind and their speed picked up making up for the lost time.

Eight tank landing craft, each carrying four DD swimming tanks, were supposed to accompany the first troops to land, but they only arrived at the transport area a half hour after they should have.

Realising that the men might have to land on a heavily defended beach without armoured support, the skipper decided to take the swimming tanks to within a short distance of the shore before discharging them. In this way they only arrived about fifteen minutes late.

The US 9th Air Force, without precision electronic bombing aids such as Gee and G-H, were to bomb the defending Germans to soften them up before the troops touched down.

Two hundred and seventy bombers of the US 9th Air Force dropped 4,500 tons of bombs, mostly some distance inland from the beach gun positions. Consequently, this huge, concentrated aerial bombardment failed to put many of the guns out of action.

The gun batteries had to be dealt with by the troops the hard way. The US navy took a hand in the gun destruction and together they more than adequately solved the problem.

When the US troops were about 2 miles off the beaches, all seventeen of the rocket launching craft launched their missiles at the German defence force.

Seventeen thousand rockets, each with the punch of a 6-inch naval shell, raised an enormous dust cloud over the sandy beaches.

From any rocket launching craft, the Naval Type 970 which was the RAF H2S radar could draw two pictures of the approaching shore. One with a range of 23 miles which allowed the skipper to see all the vessels, large and small, in the naval force off Utah beach, the other when closer to the shore in greater detail.

Once the landing craft got under way, the superb microwave radar on the rocket craft could see the individual craft loaded with troops, all in their precise positions a short distance ahead. When they reached the 4-mile mark, the skipper would switch to the 3-mile screen, enlarging the whole picture as if with a giant magnifying glass. He could then clearly see the shoreline with the LCAs approaching. When the range came down to 1,700 yards, the fixed range of the rockets, they were ignited.

With a frightening, hysterical ongoing scream, the barrage of rockets flew over the heads of the soldiers in their speeding landing craft and landed on the beach defences a short distance ahead. Little wonder that the landings were scarcely opposed.

The rockets rained death and destruction on the unfortunate German troops and raised a huge dust cloud, which blotted out the sun, shrouding the beaches in a fog. This unfortunately prevented the skippers of the approaching assault landing craft from seeing the shore. As they slowed down to get their bearings, a heavy side swell gradually moved the landing craft more than a mile south of the correct landing area.

Eventually they turned to starboard and started the run in to the beach. They found it to be virtually unprotected. Even the beach obstacles had not been mined and the men who were going to remove the mines by hand thankfully had no work to do. This was just as well as the US Command had decided that the Flail tanks, as used by the British and Canadian troops for mine disposal, were unnecessary.

The sappers on Omaha Beach were not so fortunate. Landing in the wrong place was turned to great advantage and the commanding officer studied the maps and decided that this was as good a place as any to start the war. They then prepared to negotiate the flooded countryside.

The 82nd and 101st Airborne Divisions had been practising for the drops for some time and had developed an almost pinpoint precision during exercises. They were to land near the other end of the causeways, which they were to secure and hold for their comrades landing on the beaches.

The US High Command felt that they could not rely on Air Marshal Leigh-Mallory's insistence that on D-Day, on the direct sea approach to the Normandy beaches, the German early warning radar would not work.

For this reason, the US 9th Troop Carrier Command studied the maps and completely disregarded Leigh-Mallory's orders and pleas. They did not take the planned short sea route over the Utah beachheads. Unbeknown to the commanding generals – and they could not be told – this was the route that was protected by Mandrel jamming, which provided complete protection to the British airborne troops on the remote eastern flank, by far the most hazardous of the landings.

As if on an exercise back in the United States, where the precise flight path was of little consequence, the American pilots decided to fly the long route west out towards the Bay of Biscay and reach the drop zone by overflying the Cotentin Peninsula well to the south of Cherbourg.

Mistaking Leigh-Mallory's warnings for the 'pussyfoot' British approach to fighting, which had aroused Eisenhower's wrath in the pre-Dieppe days, and despite warnings that their plan would entail flying over areas protected by dozens of German Wurzburg-controlled anti-aircraft gun batteries, the Americans decided to approach the Drop Zone from the west rather than the north.

Over the Bay of Biscay everything was peaceful and quiet, but as soon as they hit the western coast of the Cotentin Peninsula, the slow, low-flying troop transports were subjected to murderous and heavy accurate German Wurzburg-controlled anti-aircraft gun fire.

The pilots of these transport planes had no combat experience flying over the Continent, in fact many had only arrived in Europe within the last few days and had never been subjected to anti-aircraft gun fire.

As a result, large numbers of aircraft were shot down, and the thoroughly disorientated pilots dropped their troops anywhere. Most of the American airborne lads were spread over the whole French countryside, some 30 miles from their target areas.

Despite this unnecessary blunder and under impossible night conditions, a small body of the men, in true Commando style, formed up, located the causeway entrances and carried out their allotted task successfully. This was a tremendous achievement as only a sixth of the dispatched force accomplished the tasks and secured safe passage for their seaborne friends to their inland targets.

The losses, however, were grievously and unnecessarily high.

The US navy carried on off-loading vast contingents of men and materiel and the Utah landings, despite serious initial mistakes, can be claimed as a resounding success.

In military operations, little ever goes according to plan. Since time immemorial it is always the quality of the men that can turn a seeming disaster into a great victory.

This was most certainly the case on Utah.

Omaha

This, the second and major American beach, was located in between Utah and Gold.

Rear Admiral John L. Hall commanded this landing operation from his flagship, the cruiser USS *Augusta*. With him on the bridge was General Omar Bradley.

It had been decided that the six 6.1-inch guns at the Point du Hoc, together with other long-range guns reported to be in the area, posed a major hazard to his bombardment group.

The Air Force Chief, Leigh-Mallory was not held in high regard and despite the earnest and most secret assurances by him, US Admiral Hall could not be convinced. He would not believe that in the early hours of the morning of the landings, in some mysterious way German coastal defence gun-laying radar would not be working.

This, the RAF Chief insisted, applied to all surveillance radars along the 50-mile stretch of coast of the operational area of Normandy.

The reason for this apparent miracle was simply that as the ships slowly approached the coast of France, their super high-power Mandrel jammers would gradually bring down a curtain of 'mush' over all the German radar screens. By two hours before dawn this would completely obliterate all reception of signals by the German low-frequency radars, but only in the Normandy area. The area would extend from the Cotentin Peninsula to some distance east of Le Havre.

The US navy found this hard to believe and decided to ignore the suggestion and set their transport area and drop zones at a position a full 12 miles out to sea. This resulted in a rough, long voyage in to the beaches for the unfortunate assault troops.

In fact, the frequency bands used by all the Seetakt, Wasserman and Freya radars were hopelessly jammed by the super high-power

Mandrel jammers. The short-range anti-aircraft, gun-control Wurzburg radars working in the 600-megahertz band, however, were not in any way affected by the electronic onslaught.

In 1942 at Dieppe, we had travelled the same 12-mile run in to our beach and with reasonably calm seas. The two-hour sea trip does nothing to promote the morale and well-being of the men, who have to go into action as soon as they hit the beach.

At Omaha, the long journey now also required good navigation, which was almost impossible to achieve on flat-bottomed assault landing craft in conditions of very poor visibility and with no fixed navigational aids of any kind. Moreover, the longer the trip the more likely you were to run into unexpected trouble, as we had at Dieppe, when we were attacked by German E-boats.

Navigation for the skippers of the beach landing craft at Omaha was infinitely more difficult than at Utah, which lay in the lee of the Cotentin Peninsula, and was thus afforded a degree of protection from the raging wind and seas. Off Omaha, the weather was shocking, with 3- to 4-foot waves, gale-force gusting winds and driving rain. Compounding the problem was the fact that despite the pleas of naval personnel, the landing craft were severely overloaded with paraphernalia that had not been provided for in the planning calculations.

The bombardment force included the USS *Texas*, the veteran USS *Arkansas*, the cruisers HMS *Glasgow* and the Free French FFS *George Leygues*, the *Montcalm* and seven US destroyers. This was followed by the USS *Ancon*, fifteen large transports, thirty-three large infantry landing craft, and two headquarters landing craft, all escorted by five destroyers. Following in the rear was a convoy of 270 ships and general service craft, all working to a strict timetable.

Lowering the overweight landing craft caused difficulties with the davits, and this was made more difficult by the rough, uncooperative seas, which destroyed many landing craft, drowning many of their unfortunate occupants.

Off Omaha, with the winds gusting at gale force, the waves swamped many of the overloaded landing craft and it was recorded that the men were forced to use their helmets in a vain effort to bail out the sea water.

In the poor visibility it took almost an hour for the landing craft to take up some kind of cohesive formation and once they had got under way, the unfortunate men had to look forward to a two-hour trip in the bumping and bucking landing craft.

Covering the Omaha beach were sixteen well-prepared defence positions with light artillery, machine guns and anti-tank guns. The armament, mostly installed in concrete bunkers, gave the Germans complete command of the beach from a long escarpment 150 feet high. There were some breaks in the towering cliff face, but these were partially blocked with concrete, and the area had been skilfully prepared for an attack from the sea.

As on the other landings, much depended on softening up by air bombing followed immediately by the heavy pounding to be delivered by the naval bombardment group. Once under way, the LCT(R) would fire their thousands of rockets when the LCAs carrying the troops reached the appropriate range, as indicated on the microwave rangefinder. All depended on strict and orderly adherence to the approach plan that had been practised many times off the South Devon coast.

Under cover of Mandrel jamming, the bombardment ships had taken up their positions at 2.30 am and though there was plenty of time in hand, Rear Admiral Hall, for some unknown reason, had allowed only forty minutes for the naval bombardment.

On the British and Canadian beaches, the softening up barrage was to last more than two hours. Before the Germans had recovered from this stunning attack, they were to be assailed by the troops who would overrun their defences.

Everything depended on the carefully planned concentrated accurate attack, the removal of the beach obstacles, clearance of the mines and a quick turnaround of the landing craft. This had to be slickly carried out to enable the successive waves of troops and supplies to be landed quickly to overpower and overrun the defences and clear the beaches. After milling around in the darkness for some time, the assault landing craft had finally set out on their two-hour journey to the beaches.

Harried by the high wind and rough seas, they immediately lost station and in the poor light the bedraggled mass of landing craft made its bumpy way in the general direction of the still invisible beach.

Using neither Gee nor G-H, the precision bombing aids offered by the RAF, the aircraft of the US 9th Air Force dropped their lethal loads blindly through the clouds. In an effort to avoid bombing short and possibly hitting some of their own troops, the bomb aimers were told to delay thirty seconds before dropping their bombs. This resulted in most of the bombs falling 3 miles inland without in any way affecting the German gunners defending the beaches.

Perhaps a left over from General Eisenhower's attitude in 1942, when he all but accused Mountbatten of lacking moral fibre by cancelling Operation Sledgehammer, the US planners had scorned the British mode of attack instigated by the Dieppe experience, i.e., to deliberately avoid a frontal assault. Bradley had decided to take Omaha by sheer 'blood and guts'.

The US V Corps' plan rejected all use of the armoured fighting vehicles as developed by COXE, the Combined Operations Experimental Establishment, which they contemptuously referred to as the 'funnies' (in English parlance 'the comics').

Also, any thought of preparing a devious plan to approach the beaches from the flanks to avoid a frontal assault smacked too much of the carefully planned 'pussy foot' Combined Operations planning that had first aroused Eisenhower's ire in the pre-Dieppe days.

The many cogent lessons of Dieppe had been the subject of intensive study and many of the tactics shown to be vital and necessary were completely ignored. Protective weapons such as the various types of armoured fighting vehicles had been designed and manufactured as a result of protracted deliberations and experiments. Most of the new devices developed had then been operationally tested and improved during the various landings in the Mediterranean and had resulted in precision beach landings and low losses.

Sadly, the use of these wonderful new innovations to the art of amphibious war was denied to the US troops on D-Day. It seemed that another lesson was needed.

The heartless gods of war read the lesson again, at Omaha, on D-Day.

General Percy Hobart's armoured fighting vehicles were virtually a modern suit of protective armour, which gave a wonderful degree of protection to the engineers detailed to carry out the onerous task of dismantling and removing beach obstacles under enemy fire.

After the Dieppe experience, thousands of special armoured fighting vehicles of various types had been produced to protect the sappers as they went about their dangerous work on the beaches. AFVs were available to protect the men by hurling explosives at unyielding concrete strong points; for the more obstinate pillboxes, the Crocodile, a flame-throwing tank had been prepared. Special tanks were able to lay a track along the shingle over which their companions could travel without 'bellying', as had happened at Dieppe, and others could enable their friends to climb walls and overcome tank traps.

Other aids to invasion had been prepared such as the 'Flail', which literally flailed and exploded the mines in front of the tank with long iron chains. Its heavy protective armour deflected the blast of the small anti-personnel mines strewn over the beaches by the Germans.

At Omaha the US engineers had to try to clear the mines by hand and under fire with pitiful results.

On the other beaches the Flails were landed early on, and they cleared a safe passage for the troops and vehicles, speeding up the assault on the Germans still dazed by the accurate air and naval bombardment.

To achieve success the troops had to be placed on their own portion of the beach at the right time. The sea lanes between the obstacles would have to be cleared by the engineers in order that landing craft could beach, disgorge their loads and drag themselves quickly back off the beach to allow other craft to unload.

The DD swimming tanks had to land first to give cover to the troops as they stepped ashore under enemy fire. For direct support to the troops as they hit the beach, the US army was to land twenty-six heavy artillery pieces, the weapons of five artillery battalions. These had been strapped on top a small fleet of DUKW which were 2.5-ton six-wheel amphibious trucks. They had decided not to use Mountbatten's beach Landing Craft Gun, which sported two 4.7-inch guns. This stable flat-bottomed craft, the LCG, could beach and engage enemy artillery immediately.

To assist in beach clearance, sixteen armoured bulldozers were to carry out the work, which on the other beaches was done by the various specialised heavily armoured AFVs.

It seemed almost as if V Corps' Commanding General Gerow, emulating General Montgomery, was also trying to prove Mountbatten wrong, i.e., specialised equipment and training for beach landings were unnecessary and that conventional army equipment and training were more than adequate.

Omaha proved the general wrong.

The landing craft battled on through the heavy seas with the exhausted men who had not slept since the previous night. At dawn they broke their 'K' rations and tried to breakfast as their landing craft crashed from wavetop to wavetop. Understandably, many 'brought up' their breakfasts and to add to their misery, some were now ankle deep in vomit.

At least ten of the landing craft carrying troops and vehicles had already been swamped and sunk with the loss of all the men, and many others were in dire straits with the troops frantically bailing out the water with their helmets hoping and praying that they would once again get their feet back on dry land.

Generally, the landing craft drill was to approach the side of their beach in line astern coming to a start point about 2 miles offshore. They were then to turn towards the shore and the whole wave would hit the beach together at their appointed target area.

Unfortunately, all semblance of control had vanished in the long run in to the beach and there was no way of choosing the correct beach or target which the men had studied and been detailed to attack. The men were supposed to land with or shortly after the DD tanks, which were to give them covering fire once they had touched down.

Thirty-two of these tanks had been launched in the heavy seas more than 3 miles out from the shore. They had been engulfed and most of the crews drowned.

Nevertheless, in some miraculous way five DD tanks survived, but it was a case of too little too late. For the urgently needed artillery support it would seem that the men would now have to wait until the DUKWs landed their army artillery pieces.

Unfortunately, the DUKWs, hopelessly top heavy with the large guns lashed on deck, turned turtle in the heavy surf with the loss of all twenty-six large-calibre guns. This was the ultimate in disasters as there was now no artillery support for the infantry once they hit the beaches.

Dieppe had shown that at this point, the troops were at their most vulnerable. As a result, the LCT(G), a beach landing craft equipped with two massive 4.7-inch guns had been developed by Combined Operations Command. A large number of these craft was offered to Admiral Hall together with many other beach landings systems and devices, but all had been declined.

The initial force subjected to this trial by fire amounted to eight companies, or nearly 1,500 men, carried by thirty-six LCAs. Many lay prone in the sea water taking shelter behind the undestroyed beach obstacles, gradually moving forward as the tide came in.

On the run in to the beach, more large landing craft had been swamped, sinking just offshore with the loss of all the men, equipment, radios, explosives and vehicles.

Two hundred and seventy specially trained engineers were to land before the infantry. Their job was to use the explosives to destroy various beach obstacles and to dispose of the mines, making a safe path for the LCAs to beach and disgorge their troops.

The landing craft using their kedge anchors were then to winch themselves off the beach in quick time to allow other craft to beach and drop their passengers. This drill was vital if the planned amount of traffic was going to be handled in a short space of time.

Unlike on the British beaches, where AFVs protected the men as they went about their dangerous work, V Corps of the US army expected their men to carry out this difficult and spine-chilling work by hand. The engineers were to be aided by sixteen armoured bulldozers. Unfortunately, only six reached the beach and three of these were immediately put out of action by the German machine guns and mortars.

It was Dieppe all over again. Nearly half the men had been killed as they bravely but pitifully struggled with the mines and obstacles to little avail.

The monstrous pile of wrecked landing craft and materiel could not be cleared and there was now the hair-raising prospect of 25,000 men and 4,000 vehicles to be off loaded on the next tide.

Landing craft nevertheless hit the beaches, riding over the obstacles, exploding the mines and leaving a tangled mess of smashed landing craft and a beach littered with hundreds of bodies.

The landing craft continued to charge the wreckage, forcing their way through to the beach. As the ramps came down the men were instantly subjected to murderous machine-gun, mortar and artillery fire that forced the unfortunate soldiers to take refuge behind the remnants of the obstacles or any other wreckage that would give them some cover from the horrendous fire that was rained down on them by the German gunners.

With radio communications gear destroyed, there was no contact with the *Augusta* and Admiral Hall and General Bradley, fearing the worst, sent a fast motor launch to take a look at the beach area and make a report.

They were horrified as it seemed obvious that Omaha had failed with all the horrible consequences for the rest of Operation Overlord.

With men trapped on the shoreline shingle, others paralysed around the base of the escarpment and the beach piled high with the wreckage of ships, landing craft, vehicles and vast quantities of stores, it was Churchill's and Mountbatten's worst nightmare come true. If any of the standard Combined Operations craft such as the landing craft gun fitted with 4.7-inch guns had been used, or if their armoured fighting vehicles had been allowed to help the US sappers, it might have been a different story.

Furthermore, if Admiral Hall and General Bradley had taken Air Marshal Coningham's assurance at face value, which is that the German radar would not work on the day, much misery would not have befallen the unfortunate US soldiers.

It would have meant that like the British, the United States' battleships could easily have operated at 7 miles from the beaches with a comparatively short run in for the assault landing craft.

In the event, after the brave men of the 2nd US Rangers, taking heavy losses, had scaled the cliff at Pointe du Hoc and fought their way through the huge moonscape craters left by the naval gunnery, they were shocked, disappointed and surprised.

The six 6.1-inch long-range guns, which Admiral Hall had feared so much, were simply not there. Instead, six long telegraph poles had been placed in position on the concrete hard standings.

General Hobart's 'Funnies', together with Mandrel, which was so supremely successful, on that historic day could easily have saved the lives of many young Americans. The casualties on Omaha might then have been no greater than those on all the other beaches of D-Day.

That day the Allies had established a foothold in the European mainland, and though there would be many bloody battles over the ensuing months, the defeat of Germany became more of a reality. It took almost a year, for Victory in Europe was only finally achieved on 7 May 1945.

POSTSCRIPT

Dieppe

According to official figures on the **Dieppe Raid**, in addition to the **4,963** Canadian troops taking part there were 1,075 British troops (52 fatalities), 50 of the 1st US Ranger Battalion (1 fatality) and 20 of the No. 10 (Inter-Allied) Commando.

In total there were **3,367** casualties, including **1,946** prisoners of war. **916** Canadians lost their lives.

D-Day

According to the website *D-Day Revisited: The Cost of Battle,* with content written by John Phipps, on 6 June 1944, approximately **156,000** Allied troops from various countries, including the United States, the United Kingdom, and Canada, landed on the beaches of Normandy, France.

These consisted of **73,000** American (23,250 on Utah Beach, 34,250 on Omaha Beach, and 15,500 airborne troops), **61,715** British and **21,400** Canadian (24,970 on Gold Beach, 21,400 on Juno Beach, 28,845 on Sword Beach, and 7,900 airborne troops).

Initially the casualty figures were put at 10,000, with 2,500 killed. These figures have since been revised, with a total American Armed Forces death toll of **2,499** on that first day, with a further **1,915** of other allied nationalities, including mainly British.

It is now believed the British took 1,000 casualties on Sword Beach and a further 1,000 on Gold Beach. A further 700 were killed and wounded during the airborne assault. The Canadian losses on Juno Beach are recorded as 340 killed, and a further 574 wounded.

American casualties were the most severe at Omaha Beach, where it is believed the 1st & 29th Divisions suffered 2,000 casualties. By comparison, casualties at Utah Beach were relatively light at 197, including 60 missing.

Bearing in mind the total number of troops landed in Normandy that day, in percentage terms the figures are comparatively small compared to the losses at Dieppe.

ENDNOTE

The Oslo Report was written on 1 and 2 November 1939 by Hans Ferdinand Mayer, a German businessman and scientist who worked in the Siemens Telecommunications laboratory, while on a business trip to Oslo, Norway. He was opposed to the German decision to go to war and in it he described several German weapons, some already in service and some under development. He left his papers in a package which was delivered to the British legation who forwarded it to MI6.

It was considered by most of the British intelligence services to be an elaborate hoax by the German Military Intelligence Service. A copy had been given to R.V. Jones, head of the Scientific Intelligence Department, who felt that perhaps it was genuine, for with his understanding of radar technology, he was intrigued by a German radar system named Freya, the Norse goddess who could see for a hundred miles, even in the dark.

Jones collected as much information as he could and presented his findings to the Chiefs of Staff. His arguments were so compelling that finding out its full capabilities became a priority and led to Jack's participation in Operation Jubilee, often referred to as the Dieppe Raid.

APPENDIX 1

Acknowledgement of the contribution made by Combined Operations to the success of the D-Day landings – Operation Neptune.

Signal to Admiral Mountbatten.
Commander-in-Chief South East Asia Command
Ceylon India

Today on the 12th of June 1944 we visited the British and American Armies on the soil of France. We sailed through vast fleets of ships with landing craft of many types pouring more men vehicles and stores ashore. We saw clearly the manoeuvre in progress of rapid development. We have shared our secrets in common and helped each other all we could. We wish to tell you at this moment in your arduous campaign that we realised that much of this remarkable technique and therefore the success of this venture has its origin in developments effected by you and your staff of combined operations.

Signed
General Arnold
Admiral King US Navy
General Marshall
Field Marshal Sir Alan Brooke
Field Marshal J.C Smuts
Winston Churchill

APPENDIX 2

Observations made by those who took part in Operation Neptune, June 1944:

'It may probably be that weather conditions had some part in what must ever be a matter for wonder, that the embarkation, sailing and passage of the force should have been carried through without so great a movement being detected by a well-equipped, prepared and determined enemy'.

Operation Neptune, Vice Admiral B.B. Schofield (Ian Allen, 1974)

'It is difficult to understand how the great invasion armada with some 7, 000 ships and craft could sail from English ports, rendezvous in the Channel and approach the Normandy coast without detection, but this is what happened.'

Overlord, Normandy 1944, General W.G.F. Jackson. (Davis Poynter, 1978)

'Strangely, the guns of the Atlantic Wall were silent. Troops saw the coastline stretching ahead and wondered about the complete absence of enemy fire.'

The Longest Day, Cornelius Ryan, (Gollancz Wordsworth Military Library, 1960)

'But here in the narrow English Channel we could scarcely count on slipping through the enemy's alert without sounding an alarm. In some strange way we did.'

General Omar Bradley, United States Army

SUGGESTED READING

The Man Under the Radar: A Biography of Jack Nissenthall
Linda Nissen Samuels
Chiselbury Publishing, 2022

Green Beach
James Leasor
William Heinemann Ltd., 1975

Combined Operations: An Official History of Amphibious Warfare Against Hitler's Third Reich, 1940-1945
John Grehan
Frontline Books, 2022

The Rise and Fall of the Luftwaffe: The Life of Luftwaffe Marshal Erhard Milch
David Irving
Weidenfeld & Nicolson, 1974

Mountbatten: The Official Biography
Phillip Ziegler
Collins, 1985

Adolf Galland: The Authorised Biography
David Baker
Windrow & Greene, 1996

The First and the Last
Adolf Galland (translated by Mervyn Savill)
Ballantine Books Inc., 1957

The Death of the U-Boats
Edwin P. Hoyt
Warner Books, 1989

The War of the Landing Craft
Paul Lund and Harry Ludlam
New English Library, 1976

The Challenge of War: Britain's scientific and engineering contributions to World War Two
Guy Hartcup
David & Charles, 1970

German Secret Weapons (Purnell's History of the Second World War: Weapons Book No. 5)
Brian J. Ford
Macdonald & Co (Publishers) Ltd., 1970

Allied Secret Weapons (Purnell's History of the Second World War: Weapons Book No. 18)
Brian J. Ford
Macdonald & Co (Publishers) Ltd., 1971

Pre D-Day: The Allied War Machine (History of the Second World War Part 63)
Marchall Cavendish, 1974

The Secret Forces of World War II
Philip Warner
Granada Publishing, 1985

Instruments of Darkness: The History of Electronic Warfare
Alfred Price
Charles Scribner's Sons, 1978

Winning the Radar War: A Memoir
A.W. Cockerill & Jack Nissen
Macmillan of Canada, 1987

Three Steps to Victory: A Personal Account by Radar's Greatest Pioneer
Robert Watson-Watt
Odhams Press Ltd., 1957

The Birth of British Radar
Arnold 'Skip' Wilkins
Defence Electronics History Society, 2015

Echoes of War: The Story of H2S Radar
Sir Bernard Lovell
Aam Hilger, 1991

Watching the Skies
Jack Gough
HMSO, 1993

The Radar War: Germany's Pioneering Achievement 1904-1945
David Pritchard
Patrick Stephens Limited, 1989

The Invention That Changed the World
Robert Buderi
Abacus, 1998

X-Troop: The Secret Jewish Commandos of World War Two
Leah Garrett
Penguin Random House, 2021

The Commandos, 1940-1945
Charles Messenger
Grafton Books, 1991

Armes Secretes et Ouvrages Mysterieux de Dunkerque à Cherbourg
Myrone N. Cuich
Myrone Cuich, 1996

The Secret War
Brian Johnson
BBC, 1978

The Inventor of Stereo: The Life and Works of Alan Dower Blumlein
Robert Charles Alexander
Focal Press, 1999

Combined Operations, 1940-1942
HMSO, 1943

Dress Rehearsal: The Story of Dieppe
Quentin Reynolds
Random House, 1943

Operation Neptune: The Prelude to D-Day
David Wragg
The History Press, 2014

D-Day
Warren Tute, John Costello & Terry Hughes
Pan Books Ltd., 1975

Dieppe: The Dawn of Decision
Jacques Mordal
The Ryerson Press, 1962

The Shame and the Glory - Dieppe
Terence Robertson
McClelland and Stewart Ltd., 1962

One Day in August: The Untold Story Behind Canada's Tragedy at Dieppe
David O'Keefe
Icon Books Limited, 2020

Dieppe: Tragedy to Triumph
Brigadier General Denis & Shelagh Whitaker
McGraw-Hill Ryerson, 1992

From Dieppe to D-Day
John 'Jock' Hughes-Hallet
Frontline Books, 2023

Where Brave Men Fall: The Battle of Dieppe and the Espionage War Against Hitler, 1939-42
George Fosty and Darril Fosty
Stryker Indigo Publishing Company Inc., 2013

Unauthorized Action: Mountbatten and the Dieppe Raid
Brian Loring Villa
Oxford University Press, 1994

Fiasco: The Break-out of the German Battleships
John Deane Potter
Book Club Associates, 1970

Overlord: D-Day and the Battle for Normandy
Max Hastings
Pan Macmillan, 1984

Operation Neptune: The Normandy Landings, 1944
Kenneth Edwards
Fonthill Media, 2013

Bawdsey: Birth of the Beam
Gordon Kinsey
Terence Dalton Limited., 1983

Radar Days
E.G. Bowen
Adam Higler, 1987

The Radar War: Germany's Pioneering Achievement, 1904-45
David Pritchard
Patrick Stephens Limited., 1989

Britain's Shield: Radar and the Defeat of the Luftwaffe
David J. Zimmerman
Amberley Publishing, 2010

Operation Neptune: The Prelude to D-Day
David Wragg
The History Press, 2014**The route of the three German battleships during their escape from Brest, February 1942**

INDEX